Maui
How It Came to Be

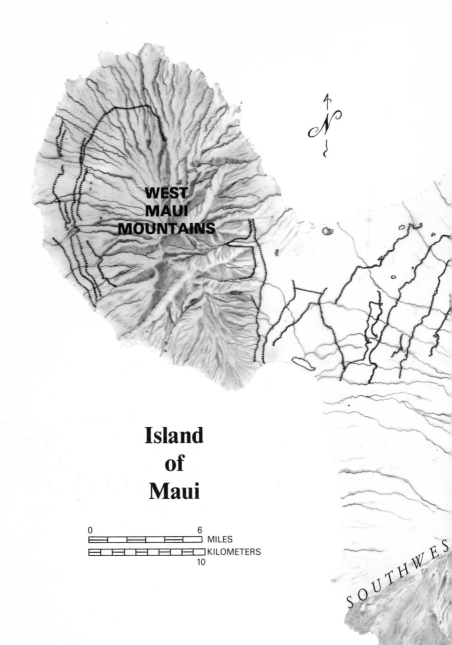

WEST
MAUI
MOUNTAINS

Island
of
Maui

N

0   6
          MILES
          KILOMETERS
       10

SOUTHWES

Maui
How It Came to Be

Will Kyselka *and*
Ray Lanterman

NORTH RIFT

KO'OLAU GAP

ALEAKALĀ

CRATER

EAST RIFT

KAUPŌ GAP

FT

THE UNIVERSITY PRESS OF HAWAII
HONOLULU

Library of Congress Cataloging in Publication Data

Kyselka, Will.
  Maui, how it came to be.

  Bibliography: p.
  1. Maui—Description and travel. 2. Geology—
Hawaii—Maui. I. Lanterman, Raymond. II. Title.
DU628.M3K9   919.69'2102   80–10743
ISBN 0–8248–0530–5

Many persons have contributed time, information, or financial assistance in the preparation and publication of this book. We thank them all:

Mary and William Aoyagi, Terence Barrow, Edward R. Bendet, Miki and Stan Blackstead, Colin Cameron, Wilson P. Cannon, Douglas B. T. Chun, Henry B. Clark, Jr., Jim Cook, Richard and Donna Corrigan, Betty Davison, A. Grove Day, Richard and Hope Depweg, Mary E. Dillon, Myrtle and Harry Endo, Kenneth Emory, Lester and Charlotte Fox.

Also, Grace Hammon, Susan and Richard Hansen, Victor Hay-Roe, Jean and Lloyd Hermel, James Y. Higa, Milton M. Howell, David and Linda Ion, Marion Kelly, Mary Kline, Timmy Leong, and Allie Lorch.

Also, George and Jo Martin, Anita McKellar, James McKellar, Art McNair, Joseph Medeiros, Edith Menrath, James and Lynda Myers, Larry and Judy Nelson, Michael and Nan Nyquist, Harriet O'Sullivan, Duane and Sarah Preble, Craig and Carol Reynolds, Loran H. Runnels, Marilyn Schulman, Neil and Helen Swanson, Myron and Laura Thompson, Clarence and Alois Vernon, and Kaupena Wong.

Generous help has also been given by Alexander & Baldwin, Ltd., and the Hawaii Science Teachers Association.

Contents

Preface vii

1 The Maui Scene 2

2 Mid-Pacific Events
and Structures 12

3 West Maui 24

4 The Isthmus 73

5 Haleakalā 83

References 153

Index 155

Seven Pools
of Pīpīwai
Stream.

Preface

Maui—How It Came to Be has been a dozen years in becoming. It began two days after Christmas 1967 when I brought a tape recorder into Gordon Macdonald's office in the Hawaii Institute of Geophysics. Earlier that year the Bishop Museum Press had published our geological history of O'ahu, *Anatomy of an Island*. Encouraged by the response to that book, we were ready for Maui.

Mac took out the geologic map of Maui which he and Harold Stearns had worked out a quarter of a century earlier. Fluently he talked out the story of a geological trip around the island. That was the beginning. Over a long period of time we added to the story and reworked it into its present form.

Our purpose was to present the geology of Maui in a clear, nontechnical way, so that the general reader could better understand the landforms of the island and the forces that shaped them. Our plan was to look first at Maui's geological history, setting, and structure. Then we'd take a closer look at its present features—its cones, craters, and caldera; its seacliffs, valleys, and coral reefs; its wave-cut benches, wind-cut terraces, and volcanic domes. But we also wanted to look at the island as human beings have interacted with it, leaving their small marks on its surface—fishponds, *heiau,* house sites, villages, towns, and irrigation ditches with their consequent greening of the Isthmus.

We wanted a presentation clear and direct, scholarly but not pedantic. And to carry the concepts clearly, we wanted a highly visual approach.

Ray Lanterman was the one to do that. He created the visuals, sometimes running time back a thousand, sometimes a million years for a look at Maui as it was during its long developing. Evolving slowly, the book became a craft operation characterized by care, fun, frustration, and an expenditure of a good deal of time. Along the way we diverted from Maui, publishing *The Hawaiian Sky, Twelve Sky Maps,* and *North Star to Southern Cross.* So you'll find here a larger perspective on Maui, with a hint of the universe at large.

Mac, in the meantime, was adding to his impressive production of more than 200 technical papers. He also published two books. One, with Agatin Abbott, was *Volcanoes in the Sea;* the other, entitled simply *Volcanoes.*

Making a book is hardly a solitary endeavor. Ray incorporated Mac's technical knowledge in his drawings of what-is and what-was, and I put the story into the written word. But behind such obvious productivity was my wife, Lee, who persistently kept Maui moving toward becoming—sometimes gingerly but always deftly.

Mac read the final manuscript. He was about to write the introduction when death overtook him on June 20, 1978. A world-renowned geologist and volcanologist, Mac is remembered by all whose lives he touched for his warmth of personality and his generosity.

Aloha, Mac, and thanks for your part in enhancing my view of the islands of Hawaii.

WILL KYSELKA

Maui
How It Came to Be

1
The Maui Scene

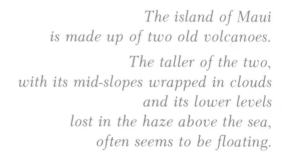

The island of Maui
is made up of two old volcanoes.

The taller of the two,
with its mid-slopes wrapped in clouds
and its lower levels
lost in the haze above the sea,
often seems to be floating.

Maui does not float, of course. It rests firmly on the floor of the Pacific 5 kilometers beneath the surface. The island that we see, gigantic as it is, is only a tiny part of the tremendous volcanic mass that exists, most of which is hidden from sight in ocean depths.

Maui lies near the southeastern end of the Hawaiian chain —a line of volcanic peaks stretching like a huge slash mark 2500 kilometers across the middle of the Pacific.

WEST MAUI MOUNTAINS   HALEAKALĀ

SEA LEVEL
OCEAN FLOOR

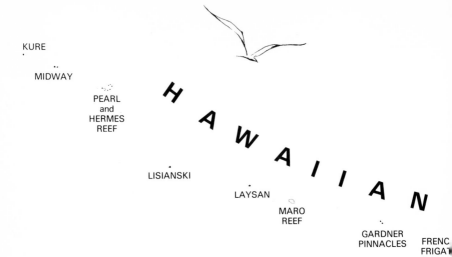

KURE

MIDWAY

PEARL
and
HERMES
REEF

LISIANSKI

LAYSAN

MARO
REEF

GARDNER
PINNACLES

FRENC
FRIGA
SHOAL

HAWAIIAN

The volcanic cones forming the islands in the chain are the most massive structures on earth. Haleakalā cone rises 8 kilometers from its base on the sea floor. Each cone in the chain is built of dark, iron-rich rock which, as highly fluid lava, poured out of vents in countless eruptions over millions of years of time.

Major eruptive centers in the chain, although separated by distances of at least 40 kilometers, are still close enough to each other so that flows from one volcano may overlap those of another. Separate cones sometimes merge in this way to form single islands.

Maui is such an island. O'ahu is, too.

O'AHU   MAUI

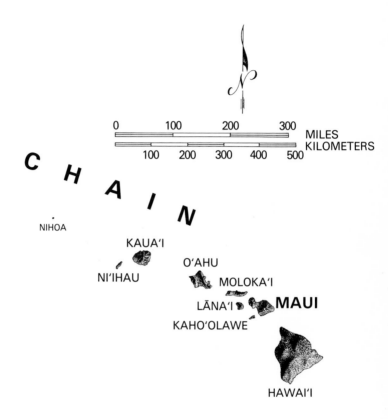

NECKER

NIHOA

C H A I N

KAUA'I

NI'IHAU

O'AHU

MOLOKA'I

LĀNA'I

MAUI

KAHO'OLAWE

HAWAI'I

0   100   200   300
MILES
KILOMETERS
100   200   300   400   500

O'AHU AND MAUI—ISLAND TWINS

A comparison of the islands of O'ahu and Maui will help us to see that the form of each follows from its structure.

Maui and O'ahu have much in common. Each is made of two volcanoes that join in a plateau. West Maui meets Haleakalā at the Isthmus; the Wai'anae and Ko'olau mountains of O'ahu meet in the Schofield Plateau.

But here the similarity ends: the two islands do not look alike. They are twins only in structure, not appearance—

fraternal, not identical, twins. For each has had a history of its own. Each has responded in its own way to the forces that shape islands. And in that response, each is unique.

O'ahu's two volcanoes are built along northwest-southeast trending rifts. Rugged, ridged, and elongated, neither has the shape we generally associate with a volcano. They are called mountain ranges, even though neither is a range in the usual sense.

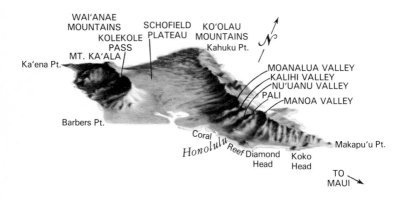

Maui's volcanoes more nearly approximate cones. One is oval in shape; the other, triangular, erupting along three fairly evenly spaced rifts.

Coral is abundant around the edges of O'ahu. Honolulu itself is, in fact, built on an old coral reef. But Maui has relatively little coral.

O'ahu has several **tuff cones** (of consolidated volcanic ash) dotting its southeastern corner: Punchbowl, Diamond Head, Koko Head, and Koko Crater. Maui's only tuff cone is crescent-shaped Molokini Island, 8 kilometers offshore in 'Alalākeiki Channel.

Tuff cones, coral reefs, and volcanic "ranges" are characteristic of O'ahu. But Maui also has features that make it unique—volcanic domes, a gigantic summit depression, and an identity reflected in its popular name, the "Valley Isle."

Diamond Head, a famous tuff cone.

Maui is slightly larger than O'ahu. If the two were perfect squares, Maui would measure 43 kilometers on a side; O'ahu, 39 kilometers.

Maui is a high island. One fourth of its surface lies above the 900-meter contour. Only a tiny portion of O'ahu, about 0.2 percent, rises that high. O'ahu, in comparison, is a low island. Half its surface lies below the 150-meter contour. If the sea should eventually reach that high stand again, as it has done in the geologic past, then O'ahu would be half its present size. Three fourths of Maui lies above the 150-meter contour.

The smaller of the two has the longer coastline. O'ahu's tidal coastline measures 330 kilometers; Maui's, 225. The shorter coastline is, as we might expect, the more rugged. Maui has 46 kilometers of sea cliff higher than 30 meters. O'ahu has only 5 kilometers of sea cliff that high.

The two islands differ in many respects, but one feature they hold in common: the most remote point from the sea on each is 16.9 kilometers.

THE MAUI GROUP

The islands of Maui, Moloka'i, Lāna'i, and Kaho'olawe form the **Maui group.** The group stands as a pedestal on the Hawaiian ridge. Kaiwi Channel, commonly called the Moloka'i Channel, separates the group from O'ahu; 'Alenuihāhā Channel separates it from the island of Hawai'i.

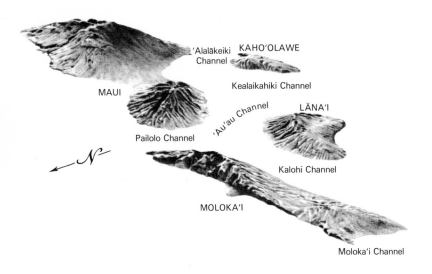

'Alalākeiki Channel
KAHO'OLAWE
MAUI
Kealaikahiki Channel
'Au'au Channel
LĀNA'I
Pailolo Channel
Kalohi Channel
MOLOKA'I
Moloka'i Channel

Rectangular Moloka'i is about 50 kilometers in length. Its straight north shore, except for the bulge at Kalaupapa, is a line of spectacular sea cliffs more than a thousand meters high. So sheer and high are the cliffs that Moloka'i looks as if it is the remaining half of an island sliced in two. Its leeward side slopes gently into the sea to meet Lāna'i beneath Kalohi Channel.

Crescent-shaped Lāna'i is a dry island, for it is in the rain shadow of Moloka'i and Maui. Huge sea cliffs on its leeward side rise 250 meters above the breaking waves.

Kaho'olawe, in the rain shadow of Haleakalā, is also a dry island. Sea cliffs on its southern side are 250 meters high. The fact that its red soil does not extend much below this level may indicate a former high stand of the sea that washed away the soil.

Sea level all over the Earth has fluctuated greatly in the last million years, as four great ice sheets moved out from polar regions, then retreated. During the periods of cold, the sea level dropped, exposing the "saddles" between the islands

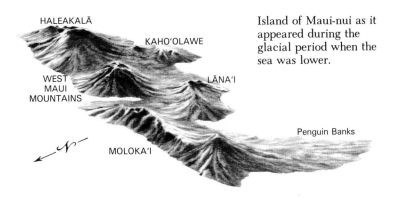

HALEAKALĀ

KAHOʻOLAWE

WEST MAUI MOUNTAINS

LĀNAʻI

MOLOKAʻI

Penguin Banks

Island of Maui-nui as it appeared during the glacial period when the sea was lower.

in the Maui group. The single large island of Maui-nui was then about half the size of the present island of Hawaiʻi.

The channels in the Maui group have Hawaiian names: Kaiwi, Kalohi, Pailolo, ʻAuʻau, ʻAlalākeiki, ʻAlenuihāhā, and Kealaikahiki. A suggestion of more than casual communica-

tion between Hawai'i and Tahiti is contained in the last name, for in Hawaiian it is literally 'the path to Tahiti'.

Chants, stories, and legends tell of repeated voyages between Hawai'i and Tahiti. Such journeys must represent the ultimate in the ability of Polynesians to sail long distances to remote islands and return—all without the use of navigation instruments. We can imagine, then, Kealaikahiki Channel giving direction to the ancient voyagers and Haleakalā serving as a landmark by day, until, beyond sight of the peak, they turned to the stars as guides.

THE ISLAND OF MAUI

Maui has the shape of a figure eight, with two large bays at its waist. The low, broad Isthmus (often called the Central Plain) lies between Kahului Bay at the north and Mā'alaea Bay at the south.

The island's two volcanic cones differ greatly from each other. One is half as high above the sea as the other. One is heavily dissected; the other, comparatively smooth. One is

HALEAKALĀ, A YOUNG CONE

Usual Height of
Trade Wind Clouds

Mā'alaea Bay

Kahului Bay

WEST MAUI MOUNTAINS,
AN OLDER CONE

Island of Maui as approached from O'ahu.

completely covered with vegetation; the other, bare rock at high altitude. One is extinct; the other was active only two centuries ago.

Now with a general view of Maui in mind, we'll have a look at how it came to be—from its early origin in deep ocean waters to its present period of shaping in the atmosphere.

2
Mid-Pacific Events and Structures

Far below the waves of the Pacific Ocean,
Earth's crust,
breaking in response
to internal stresses and strains,
opens channels to chambers
of molten magma.

Welling up through cracks and fissures,
incandescent lava spreads
out onto the ocean floor—
white molten rock,
sliding into the frigid darkness
at the bottom of the sea.

Yet nothing appears at the surface. Not a glow, not a glimmer, not a ripple; not even the slightest change in the ocean's temperature indicates the magnitude of the event taking place so far beneath its surface. The birth of a sea-floor volcano is a hidden event—hidden by deep water which can transmit the news only a short distance.

Water has a high heat capacity, taking on great quantities of heat with only a slight change in its temperature. So it readily absorbs the heat of an undersea eruption, dispersing it far and wide in deep ocean currents. Water also absorbs light. The brilliance of an ocean-floor eruption is quickly swallowed up in the opaqueness of the water. Because of this property of water to absorb both light and heat, an eruption must be within a few hundred meters of the surface to be detected.

The composition of lava erupting on the floor of the sea is the same as that of lava erupting on land. But the two environments are vastly different. We look for that difference to be expressed in the rock which forms.

13

Lava spreading out onto the ocean floor beneath 5 kilometers of water is under great pressure. Gases within it cannot readily expand, and so the rock that forms is dense and solid. On the other hand, lava flooding out onto land encounters only the relatively low pressure of the atmosphere. Gases within it expand easily, forming bubbles similar to bubbles that form in baking bread. Rock filled with bubble holes is popularly know as puka-puka rock.

The birth of Maui. The volcanoes of the Maui group appeared at different times, but each may have looked as pictured here, as it emerged from the sea.

KAHO'OLAWE

HALEAKALĀ

PROFILE OF GIANTS

Volcanoes along the Hawaiian chain
are rounded, dome-shaped structures,
broad for their height.

Resembling, in profile, the shields
of medieval warriors, the huge cones
are known as **shield volcanoes.**

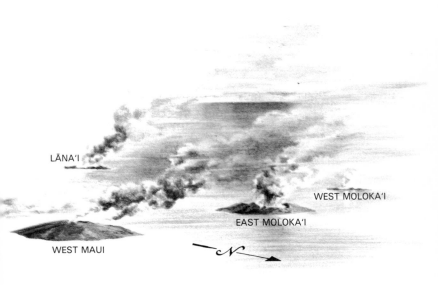

LĀNA'I

WEST MOLOKA'I

EAST MOLOKA'I

WEST MAUI

Shield volcanoes are made of an accumulation of very mo-
bile lava. Erupting at high temperature, the lava runs swiftly
from the vent, spreading widely, adding to a cone of gentle
slope. A thousand times more viscous than water, the lava
may still reach speeds of 40 kilometers per hour in narrow
channels and travel as much as 50 kilometers before solidi-
fying.

15

Hawaiian volcanoes generally erupt along cracks in the cone's flanks. Movement within the Earth's crust tears open a whole series of fissures, forming a **rift zone.** Squeezing up through the cracks, magma fills the fissures and spills out onto the land as lava.

A shield volcano may be nearly circular when viewed from above (like West Maui); or, if it grows along prominent rift zones (as in the Koʻolau volcano of Oʻahu), it may become a long oval, resembling in form an upside-down canoe.

When an eruption is over, the lava remaining in the fissure hardens into a wall-like mass called a **dike.** Generally, dikes are about a half meter in width and may extend hundreds of meters in length. Prominent dikes appear in Ukumehame Canyon on West Maui and in the summit depression of Haleakalā. Where streams have cut canyons in old rift zones, we can sometimes count as many as 400 dikes in 1 kilometer of distance.

Shield volcanoes are prodigious lava producers. Haleakalā alone contains enough material to make a hundred mountains the size of Fujiyama in Japan. Yet for all its height, bulk, and mass, it is not the shield volcano but the **composite volcano** that is so picturesque.

The classic upswept form of the composite cone is seen in many of the world's famous mountains—Fuji, Mayon in the Philippines, Hood, Rainier, Shasta, and St. Helens in the Pacific Northwest, and Vesuvius in Italy.

Composite cones are found around the edge of the Pacific but not within the basin. Made of lava of low mobility, the composite volcano alternates soft beds of cinder, ash, and fragmental material. Highly viscous, the lava does not flow

10,000-FT. SHIELD VOLCANO

SEA LEVEL

OCEAN FLOOR

MAGMA RISES IN FISSURES TO FOUNTAIN AS LAVA

ERODED DIKE

COOLED LAVA IN OLDER FISSURE
FORMS A DIKE
(Shown exposed by erosion of
surrounding terrain)

RESERVOIR
OF
MAGMA

Cut-away diagram of
volcano's flank showing how
lava dikes are formed.

Although approximately the same
height above sea level, Hawai'i's
shield volcanoes, rising from the ocean
floor, are very much larger than
composite volcanoes.

10,000-FT COMPOSITE VOLCANO

far from the eruptive site but piles up around the vent, building it higher and higher. Cinder, by its very nature, builds a loose, weak structure that slumps under its own weight and is easily eroded. But the interbedding ribs of lava in the composite cone strengthen the structure, allowing the cone to attain great height and grace.

The sides of a composite volcano steepen toward the top, and the view opens up as you climb. At the peak, the feeling of height and isolation pervades. But climbing a shield volcano is an experience of quite a different sort. The convexity of the shield itself limits the view; at the summit you are aware of the massiveness of the mountain, not its towering isolation.

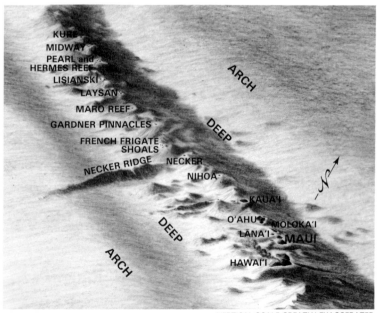

VERTICAL SCALE GREATLY EXAGGERATED

The Hawaiian Ridge rising from the floor of the Pacific.

HOT SPOT

For perhaps a hundred million years, the plate of Earth's crust beneath the Pacific Ocean has been moving over a hot spot in the mantle below. During that time, the hot spot has remained stationary, just where it is now beneath the island of Hawai'i. Time and time again, molten rock has risen from the hot spot to build volcanic islands which have then drifted away due to the movement of the plate.

The oldest volcanoes created by the hot spot are the submerged Emperor Seamounts. At the time of their formation, the Pacific plate was moving northward. But about forty million years ago, the plate received a nudge to the northwest, and the Hawaiian Ridge came into being. As a result of the change in direction, the Emperor Seamounts now lie to the north of the island of Midway.

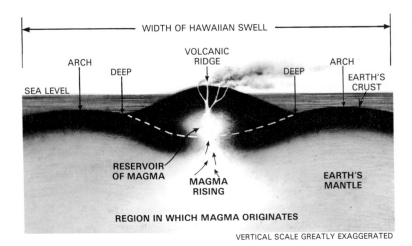

Cross section of the Hawaiian Swell showing downbowing of Earth's crust beneath the ridge.

GEOLOGICAL HISTORY OF MAUI

Outpourings of lava at the southeastern end of the Hawaiian Ridge built two gigantic shields that merged to become the island of Maui. The West Maui volcano was the first to reach the surface of the ocean. It may have risen 1 or 2 kilometers above the sea before Haleakalā appeared as a tiny islet 40 kilometers to the southeast.

We recognize, in both West Maui and Haleakalā, three periods of volcanic activity, identified by rocks of three different **volcanic series.**

WEST MAUI VOLCANO

Thin flows of iron-rich rock (basalt) built the West Maui shield to a height of more than 6 kilometers above the ocean floor. Rocks of this shield-building period are assigned to the **Wailuku Volcanic Series.** Near the end of the period, pressure beneath the summit decreased, and the top of the volcano collapsed to form a sunken crater, or **caldera,** about 4 kilometers across. Lava flooded out onto the caldera floor, and repeated collapse fractured the horizontal lava layers.

A short rest period followed the completion of the Wailuku shield, long enough for soil layers to develop. Activity began again, but this time the lava was of a different sort (mugearite and trachyte)—a lava lighter in color than the iron-rich basalt and more closely related to granite in composition. Lava of this **Honolua Volcanic Series** forms a thin whitish veneer over the darker rocks of the Wailuku series.

A long period of erosion followed the Honolua activity. Streams cut deep canyons in the shield while wave action carved cliffs at the water's edge. Sea level rose and fell many times, alternately joining and separating West Maui and Haleakalā. Renewed, weak volcanic activity built four cones of silica-poor basalt (basanite) near the sea. Rocks of this period belong to the **Lahaina Volcanic Series.**

EMERGENCE

WAILUKU VOLCANIC SERIES

CALDERA

HONOLUA VOLCANIC SERIES

KAHAKULOA

LAHAINA VOLCANIC SERIES

WEST MAUI TODAY

←(HALEAKALĀ)

KAʻANAPALI   LAHAINA   OLOWALU

All views are from the northwest.

21

HALEAKALĀ VOLCANO

Haleakalā originally built a shield of basalt about 7500 meters above the ocean floor. Rocks of this shield-building period belong to the **Honomanū Volcanic Series.** The shield is made of flows of rough, clinkery-surfaced **aa** with dense interiors in beds about 5 meters thick, alternating with thin beds of smooth, ropy-surfaced **pahoehoe.** Honomanū flows ran down Haleakalā, ponding against West Maui at the Isthmus.

The frequency of eruptions decreased as the volcano reached maturity, and the chemical composition of the lava changed. Big, bulky cinder cones formed along rifts that extended eastward, southwestward, and northward from the summit. Flows of this **Kula Volcanic Series** mantled the Honomanū shield to a thickness of 800 meters near the summit, but only one tenth that thickness near the sea.

A long period of erosion followed the veneering of the shield, just as it did on West Maui. Canyons, 1500 meters deep, were cut but later filled by the fluid flows of the **Hāna Volcanic Series.** Cones of the series stretch from Hāna up the east rift, across the summit, and down the southwest rift to La Pérouse Bay.

All views are from the south.

3
West Maui

West Maui is a shield volcano roughly circular in shape, slightly elongated in a north-south direction.

Volcanic activity built the shield; running water shaped it.

Now, with thick vegetation covering sharp ridges and filling deep canyons, the deeply eroded shield is known locally as the West Maui Mountains.

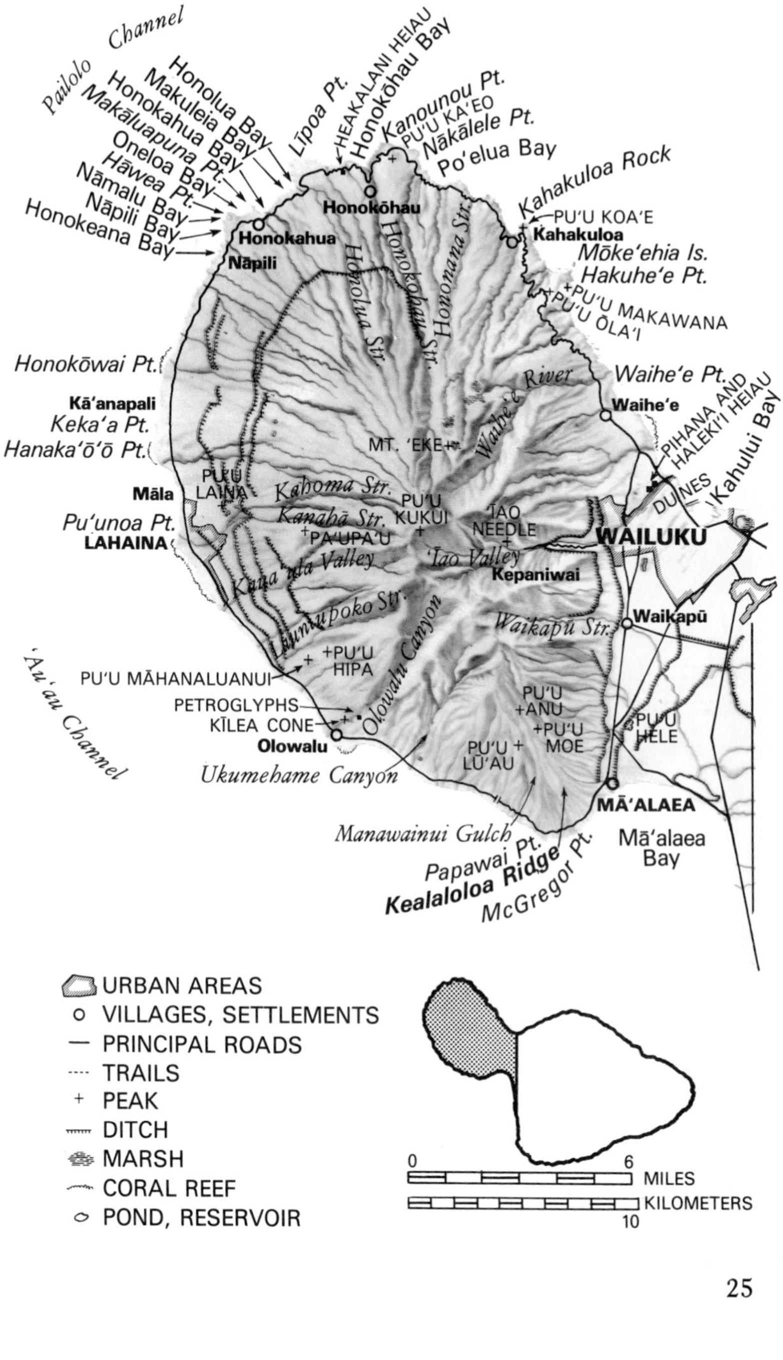
Pailolo Channel
Honolua Bay
Makuleia Bay
Honokahua Bay
Makāluapuna Pt.
Oneloa Bay
Hāwea Pt.
Nāmalu Bay
Nāpili Bay
Honokeana Bay
Līpoa Pt.
HEAKALANI HEIAU
Honokōhau Bay
Kanounou Pt.
PU'U KA'EO
Nākālele Pt.
Po'elua Bay
Kahakuloa Rock
PU'U KOA'E
Kahakuloa
Mōke'ehia Is.
Hakuhe'e Pt.
PU'U MAKAWANA
PU'U ŌLA'I
Honokōhau
Honokahua
Nāpili
Honolua Str.
Honokōhau Str.
Honanana Str.
Honokōwai Pt.
Kā'anapali
Keka'a Pt.
Hanaka'ō'ō Pt.
Waihe'e Pt.
Waihe'e
Waihe'e River
PIHANA AND HALEKI'I HEIAU
Kahului Bay
DUNES
MT. 'EKE
PU'U LAINA
Māla
Kahoma Str.
Kanahā Str.
PU'U KUKUI
'ĪAO NEEDLE
Pu'unoa Pt.
LAHAINA
PA'UPA'U
WAILUKU
Kaua'ula Valley
'Īao Valley
Kepaniwai
Launiupoko Str.
Waikapū Str.
Waikapū
Olowalu Canyon
'Au'au Channel
PU'U MĀHANALUANUI
PU'U HIPA
PETROGLYPHS
KĪLEA CONE
Olowalu
PU'U ANU
PU'U MOE
PU'U LŪ'AU
PU'U HELE
Ukumehame Canyon
MĀ'ALAEA
Manawainui Gulch
Papawai Pt.
Kealaloloa Ridge
McGregor Pt.
Mā'alaea Bay
URBAN AREAS
VILLAGES, SETTLEMENTS
PRINCIPAL ROADS
TRAILS
PEAK
DITCH
MARSH
CORAL REEF
POND, RESERVOIR
0
6
MILES
KILOMETERS
10

TWO VIEWS FROM KAHULUI

Kanahā Pond and the Kahului breakwater are two vantage points for viewing the West Maui Mountains.

The calm waters of Kanahā Pond reflect the beauty of the old volcanic cone; imaged in tranquility, the ancient peak seems timeless and eternally serene. *Kanahā* means 'shattering', but the only shattering that takes place in this wildlife sanctuary is that momentary dissolving of the inverted mountain image as birds land on the still water.

The view from the end of the Kahului breakwater is almost the same, but the scene is dynamic. Waves pound the breakwater, splash over it, and ruffle the waters behind. The har-

'Īao Valley as seen from Kanahā Pond.

West Maui, from McGregor Point to Hakuheʻe Point, as seen from Kahului breakwater. It is easy to imagine the low profile of the original ancient shield, now heavily eroded.

bor is much too rough for reflection, too restless for the mirroring of mountain images. But in that movement we're in touch with the power that neither manmade walls nor mountains can long endure.

Huge squared-off chunks of basalt, smoothly faced and as tall as a person, form the 700-meter-long walls of the breakwater. How solid and ponderous these 20-ton boulders are to our step, and yet how delicate are the tiny cubical crystals of salt that form in the shallow depressions on their surfaces! Drill holes in the boulders, 5 centimeters in diameter, remind us of the power it takes to wrest these rocks from the dense interiors of old lava flows.

Y-shaped concrete castings at the end of the breakwater look like gigantic jacks tossed randomly into the sea. White and glaring in the sun, regular and rigid in form, they are unnatural in this setting; but at the water's edge they're pink and black with algae. The sea sloshes back and forth in the spaces between the jacks in a sweeping motion that livens the circulation within the harbor.

Long ocean swells passing through the narrow opening in the breakwater decrease in wavelength, lose energy, and lap quietly upon the shore—a wave diffraction pattern clearly seen from the air. Behind the barrier, the red and black channel buoys nod and bob to a rhythm different from the rhythm of those that ride the open sea.

Early in the morning, with the sun coming up behind Haleakalā, the clouds above West Maui take on pink and yellow colors, a pattern imitating the shape of the land below. The day warms, the veil thins, and then you can see through the gap in the crest of the mountain made by ʻĪao Stream. Valleys steep and deep appear and disappear as clouds move in and out. The angle of the sun changes with the hour, modeling the relief.

Sunlit Waiheʻe and Paukūkalo sand dunes stand out against the gray-blue distant slopes. On another sand dune above ʻĪao Stream are the foundations of two important ancient Hawaiian temples—the *heiau* Halekiʻi and Pihana.

We can look far out over the ocean and imagine the excitement among the people here two centuries ago as the great sails of Captain Cook's ships came into view on the horizon. Earlier that year, on January 18, 1778, the Cook expedition sighted Oʻahu and landed on Kauaʻi. Now they were returning from a summer's exploration in Alaska where they had searched for the Northwest Passage. Sailing into Kahului Bay in November, they stayed well offshore and did not land, for the big right-angled bay is too open to provide much shelter. They cruised windward Maui and Hawaiʻi for the next six weeks before finding a suitable harbor on the leeward side of Hawaiʻi at Kealakekua Bay.

Eleven kilometers north of the Kahului breakwater is Hakuheʻe Point. It is a blunt, blocky wall of white lava against which the sea breaks. Perched on that wall 300 meters above the waves is the cone Puʻu Ōlaʻi. Farther inland, the long line of the West Maui shield rises to its 1754-meter summit, Puʻu Kukui, then slopes gradually toward the south to enter the sea in the low cliffs of McGregor Point.

Waiheʻe, Waiehu, ʻIao, and Waikapū streams have cut deeply into the West Maui volcano, transporting its core material and depositing it in huge fan-shaped banks—**alluvial fans**—at the mouth of canyons. Although separated from each other by distances of 3 kilometers, the canyons are still close enough together so that their fans have joined to form an **alluvial plain** 13 kilometers long. This apron of alluvium is clearly seen where it has been planted in sugarcane.

Wailuku town rests on that transported core of the West Maui volcano. Dominating the town a century ago were the Kaʻahumanu Church steeple and the smokestack of the sugar mill. Wooden buildings with high false fronts lined Main Street, and the town had a sharp edge, stopping where the canefields began. But high concrete structures now tower above old buidings, and the town hitches upward on the fan.

The road into ʻIao Valley passes the museum of the Maui Historical Society—Hale-hōʻikeʻike, 'house of display'. Huge trees shade the yard and a large green lawn gives space to the

130-year-old rambling white house. To enter is to take a step into the past and to feel what it was like when kitchens had dirt floors. Ancient Hawaiian artifacts as well as century-old implements show man's ingenuity in devising tools to shape his environment. Walls of this house, built in 1841 for the Reverend Edward Bailey, are 30-centimeter blocks of coral covered with plaster. Human hair, used as the binder in the plaster, was donated by the young women who attended the Reverend Bailey's seminary.

'ĪAO VALLEY

The road enters 'Īao Canyon at an elevation of 300 meters. Strong evidence for the shifting seas of glacial times is found here in two terraces. The higher terrace was formed at a time when the sea stood about 30 meters higher than it does today. 'Īao Stream then flowed across the alluvial plain and into the ocean. But as worldwide climatic conditions changed, sea level began dropping and 'Īao Stream dug deeper into its own alluvial deposits. Once again, sea level dropped and the stream further incised the fan, leaving the second, lower terrace.

'Īao Valley widens at Kepaniwai, and 'Īao Needle is slightly more than a kilometer farther on. Only by walking that short distance can you begin to know the grandeur of the valley and to sense the magnitude of force and time needed to carve it. Side canyons, hidden from each other by sharp fin-like divides, contain waterfalls in their dark recesses. Hanging high above the valley floor are streams that end abruptly. Water tumbles from these truncated channels into plunge pools; sometimes it is blown into a spray in which the sun makes rainbows.

Kepaniwai is the site of a battle between two powerful chiefs in 1790. Kalanikupule was ruling Maui while his father, Chief Kahekili, was on O'ahu attempting to gain control of all the islands. Kahekili was well on his way to uniting the islands under one command. But the young Kameha-

The two terraces in ʻĪao Valley. At the present time, streets and buildings of Wailuku cover much of the lower portion of this view.

meha of Hawaiʻi was also on his own rise to power. Landing on Maui, Kamehameha defeated Kalanikupule at Kepaniwai with the help of guns from another culture.

Kalanikupule escaped by climbing the trail at the end of ʻĪao Valley, then descending the other side into Olowalu. The two chiefs were destined to meet again, this time at Nuʻuanu Pali on the island of Oʻahu, where Kalanikupule was once again defeated.

Banks of alluvium along the side of the road contain large angular blocks of rock scarcely weathered at all. But with them also are greatly weathered spheroidal rocks, stained a rusty color, that rest on decomposed rock as soft as clay. A layer of soil covers it all to a depth of 5 meters.

It takes a long time to carve a canyon such as 'Īao—ten thousand human lifetimes . . . maybe twenty thousand. Using pebbles and boulders as tools, and with time in its favor, the stream swirls stone against stone, grinding off points, chipping edges, and polishing rounded surfaces. Big pieces of rock become smaller ones in Nature's abrasion mill, and the products are fragments of the parent rock.

A stream carries tiny particles in suspension, but larger ones leap along the stream bed in a jumping action. Cobbles and boulders tumble and roll along the bed with a rumble like that of distant thunder. Air and earth vibrate as a mountain moves toward the sea in flood time.

Yet for all its power to break rock, a stream is chiefly a transporter of finer material already broken up in chemical and biological processes. Rock is altered in the process of chemical **weathering** so that the end products are very different from the parent rock. Decomposition begins on all rock surfaces. Water and air penetrate the rock mass through fractures, cracks, and jointing planes. Some minerals absorb water, swell up, and crumble; others oxidize. Weak places are the first to go, and the surface becomes irregular.

Chemical elements in the minerals may combine with oxygen to form oxides, or they may join with material already in solution to form carbonates and clay. Lichens and algae contribute to rock rotting. They live in pits about 0.2 millimeters in diameter, and the rock material removed is probably absorbed by the plant.

Decomposition speeds up if waters contain soil bacteria. Dead organic matter adds carbon dioxide to circulating waters, forming carbonic acid. Bacterial fermentation also produces acids that further increase the solvent power of water. In each block of rock decomposition works inward toward a spheroidal core.

The rate of decomposition depends upon the surface area of the rock. Particles greater than 1/16 millimeter in diameter are classified as **sand**. Smaller particles are **silt**. Tiniest of all are **clay** particles—less than 1/256 millimeters—so small that 100 can fit across the period at the end of this sentence. A stream can carry much of its load as invisible particles. Half the load of great slow-moving rivers may be composed of particles of silt.

Sunlight reflects from the surface of 'Iao Stream while lambent shadows move across its graveled bed. Funneling through the narrow spaces between wedged-in boulders, the water humps up into a standing wave, dives under, and releases energy in a froth. An active bubble raft, formed at a place of upwelling, moves downstream—a delicate platform continually reshaping and disintegrating as it travels.

Time and flow. Hour after hour, the stream rolls on. It rains . . . a misty rain that cools the sun-heated boulders, making not the slightest change in the volume of 'Iao Stream. It's the heavy rains high up in the valley that add to its volume. After hours of downpour in the higher regions, the stream is a cataract of white water, its power to transport material increasing in geometric proportion to its volume.

'Iao Canyon broadens at its head into a natural amphitheater 3 kilometers in diameter and about 1000 meters deep. The valley floor is cool, dark, and serene in the morning. Sunlight touching the tops of ridges reveals a series of partitions in infinite regression, each a little more misty, each a little more mysterious.

This place of present quietness was once the violent center of the West Maui volcano. For millions of years, molten rock rose to the surface throughout this region, fountained into the air, and ran down the sides of the mountain. Time and time again, the withdrawal of magma from beneath the summit resulted in collapse that formed a caldera a few kilometers in diameter. But the repeated return of the magma increased the summit pressure, wrenched rocks apart, and flooded the floor with new flows. Within the caldera, the rocks lie flat; outside, the old flows slope 10 to 20 degrees toward the sea— a much steeper slope than is usual in a shield volcano. Conduits to old magma chambers show as long streaks of dense rock—dikes—stretching from floor to summit in 'Iao Valley.

'Īao Stream.

Stages in the formation of a valley such as 'Īao. *Upper left:* Summit eruption on a shield volcano; a crater is formed. *Upper right:* As activity subsides the crater floor sinks, creating a caldera. Erosion begins; gullies form on the mountain's flanks. *Lower left:* Erosion continues; a major stream eats into soft material of the caldera and begins to form an amphitheater. *Right:* The stream has eroded the caldera and the amphitheater is completed. Other streams are slowly working their way headward.

Hot, steamy gases rich in sulfur rise to the surface for thousands of years after the outpouring of lava ceases, condensing against rock walls to form crystals. The minerals in the rock through which the gases move are altered. Soft chlorite replaces pyroxene, soft clay replaces hard feldspar, and more silica is released in the process. Now in solution, silica is free to move through cracks and crevices. It is redeposited as opal and chalcedony "moonstones" that you find in 'Īao Stream, washed out from the caldera region.

'Īao was the first stream, and so far the only one, to reach the old eruptive center of the West Maui volcano. Gas-softened rocks erode easily. Once 'Īao Stream tapped the cal-

ʻĪao Needle.

dera, the stream's waters scoured it out quickly, transporting the material out of the valley, onto the Isthmus, and some of it into the ocean.

The point of 'Īao Needle is 685.5 meters above sea level; it rises about 400 meters above the valley floor. Sharp, pointed, and surrounded by deep V-shaped canyons, it appears to be an isolated pinnacle, justifying its name. But if you hike to the end of the trail past the observation shed, you will see that, rather than being needle shaped, it is more a knob on the end of a long sharp ridge. 'Īao Needle is a resistant chunk of caldera fill, strengthened by a large number of dikes. Its Hawaiian name, Kūkae-moku, is translated, with a measure of reserve, as 'broken excreta'.

The trail beyond the observation point runs along the crest of an alluvial ridge between two tributary channels. Beside it grow tree fern, ti, 'ōhi'a lehua, lantana, begonia, and orchid. Vegetation is thick and the region is moist, so the hike is often a wet one. Clumps of fern wedged into cracks add a bit of green to the massive black rock. And birds, flying in the sunlight in front of shadowed amphitheater walls, lend scale where magnitude is hard to comprehend.

Clouds pour over the 1754-meter summit of West Maui and spread out into the valley. But as the day warms, the clouds absorb heat and evaporate—a cloud-tumbling act with disappearing clouds! When equilibrium is reached, a steady line of cloud is formed, effectively drawing a curtain around the peak. The name, 'Īao, means 'cloud supreme'.

WAIKAPŪ

We leave 'Īao Valley and travel southward across the alluvial plain, past a 3-kilometer stretch of crenulated mountain slope, to Waikapū, 'water of the conch'. The stream of Waikapū has worked its way far headward and has reached caldera rocks, but at too high a level to drain the caldera; the master stream, 'Īao, does that. Sharp, steep ridges and gullies on alternate sides of Waikapū Valley look like huge interlac-

Upper Waikapū Valley.

ing fingers, the stream weaving back and forth at the intersections.

Flowing out of its canyon and through the little village famous in song as the home of "Maui Girl," Waikapū Stream disappears into the Isthmus. Thousands of years ago, when the sea was lower, it flowed northward and entered the ocean at Kahului Bay. But sand dunes piled up by the wind blocked that route, forcing the stream into Māʻalaea Bay. The stream, however, no longer reaches the sea: most of the water is taken out for irrigation, and the rest merely sinks into the ground.

PUʻU HELE

Late in the history of West Maui, the ground cracked near Māʻalaea. Molten rock charged with gas squirted into the air to build the cinder cone, Puʻu Hele. A cinder cone is built by lava fountaining from a vent. Blobs of lava spin through the air, twisting and turning, forming **lava bombs** with curlicue ends. Material propelled with less force may merely rise into the air and flop down near the vent as **clots** and **spatter**. Cinder is porous material with a burned-out look. When so full of holes that it is light and cellular, it may even float on water as **pumice.**

Pu'u Hele was once a cinder cone 20 meters in height. Now it's a reddish brown hole in the ground, mined for its cinder content; it is twice as deep as it once was high. Because cinder makes good road bed material, Pu'u Hele is spread over a sizeable portion of Maui.

To descend into Pu'u Hele is to take a trip back into time, for history is written in the walls. At the top is a 3-meter layer of brown soil, grading downward into black cinder. Clumps of cinders stucco the walls. The stratification is due to pulsations in activity during the eruption as well as changes in wind direction. There is a feeling of the ancient and primitive in this old eruptive center that dwarfs a solitary person.

Once a cinder cone, Pu'u Hele today is a pit 40 meters deep, quarried for its excellent road-building material. The history of the cinder cone is recorded in the stratified walls of the pit. Curved vertical marks are claw marks made by earth-digging machines.

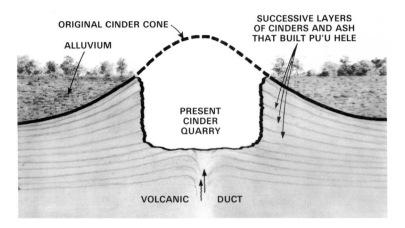

Diagram of Pu'u Hele.

McGREGOR POINT

McGregor Point is a prominent headland 250 meters wide and 30 to 60 meters high. It was built by later lava flows running down the mountain and pouring into the sea. Waves batter this peninsula during **kona storms** when winds blow from the south.

A cove on the west side of McGregor Point provides a somewhat protected place for a boat landing. But it must have taken a good deal of courage as well as skill to tie up here even at the best of times, for the water is rough. Three deeply buried concrete pillars and a boat mooring clamp thick with rust are all that remains of McGregor's Landing.

A trail begins above the cove, runs along the edge of a lava flow and down to the sea. One side of the trail is a wall of dense lava; the other, a collection of angular fragments split off from that wall. Beneath the dense flow lies a layer of soft rock. Easily scratched with the thumbnail, the rock erodes rapidly in some places, into rounded knobs; in others, into deep pits where spiders weave webs and wasps build nests. Undermining of the massive flow has created open caves and

Overhanging cliffs near McGregor Point.

overhanging ledges. No longer supported, the rock splits off in angular chunks along jointing planes and tumbles to the ground beside the trail.

The McGregor Point flow is a massive bit of natural sculpture which preserves the ancient form of the viscous lava as it slid beneath the hardening crust—movement frozen in rock. More of the action becomes evident as erosion reveals it.

The trail descends the sea cliff and opens onto a rock bench cut by a higher sea. Here tide pools, a meter or two in diameter and half as deep, fill and empty with the rhythm of the waves.

Each pool is the home of a variety of living things—hunters, grazers, and sitters. It is the home of the hermit crab with his awkward gait and whimsical appearance in his borrowed shell; the black grapsid crab that scurries about the rock walls and into crevices when he senses someone near; and the temporary home of a fish caught in a tidal change who wiggles his tail and waits for a surge of the sea to skip to another pool. It is the home of periwinkle and limpet, starfish and sea urchin, ʻopihi and barnacle, gray and brown algae, sponge, coral, seaweed; and animals too small to be seen with the unaided eye.

Torrents sweep in and out of the spaces between rocks, and water cascades from one pool to another. Responding to the tug of the moon, the sea retreats for a few hours, leaving the pool isolated and quiet, with only sunlight and wind touching its surface. The hot sun increases the temperature of the small pool; evaporation makes it saltier—rapidly changing conditions to which only the hardiest organisms have adapted.

Life hangs on tightly at this interface of land, sea, and air. Here we're in touch with the whole succession of life as it developed on this planet—the invertebrate, the fish, amphibian, reptile, bird, and mammal. We ourselves are but detached packets of ocean water, carrying in our blood the same salt content as the sea around us. And we also anticipate the occasional high-peaking wave, stepping back from it perhaps

Tidal pools at McGregor Point.

because of a sense born of thousands of generations of humans who lived beside the sea and knew its ways.

The sea surges against this point of land today just as it has done in the past and just as it will do for thousands of years to come, with scarcely a change. Changes there are, of course, but over a magnitude of time that eludes us. The ocean will retreat as another Ice Age approaches. Waves will break farther from old shores. Tide pools, then high and dry, will be filled only by occasional rains. Winds will sweep over old coral heads. Cliffs will stand far back from the waves for a hundred thousand years until a warming period will bring the sea back, once more to break against these rocks.

LEEWARD WEST MAUI

Leeward West Maui is hot and dry at sea level. The desert brown colors of the lower slopes give way to the greens of rain forests as the grade steepens. The gulches are wide, and stubby lava flows dip into the sea like toes. As we travel toward the north in a great unrolling curve, we move from a region of rugged sea cliffs to one where gentle slopes enter the sea.

Behind McGregor Point is Kealaloloa Ridge. Lying along the ridge are three **volcanic domes**—mounds of viscous lava built over volcanic vents. Puʻu Lūʻau, at an elevation of 712 meters, lies just west of Manawainui Gulch. Northeast of it and 30 meters higher is Puʻu Moe. The highest of the three, 906-meter Puʻu Anu, lies 5.5 kilometers from the sea. The slope of Kealaloloa Ridge is 9 degrees, although it seems much steeper.

Manawainui Gulch is 2 kilometers beyond McGregor Point, just past Papawai Point. The big, deep gulch is the boundary line between the Wailuku and Lahaina districts. It is an unusual dividing line, for ridges, not gulches, are the customary political boundaries. At this dry southern end of West Maui, however, the ridges are broad and flat, and the gulches which cut into them are prominent. Wide and can-

Andesite flows, more viscous than fluid basalt, formed steep, stubby slopes just west of McGregor Point.

Kealaloloa Ridge as seen from Keālia Pond. The three peaks, in ascending order, are the trachyte cones of Puʻu Lūʻau, Puʻu Moe, and Puʻu Anu.

yonlike at the sea, the gulches pinch out with altitude; at 300 meters, they disappear.

A 3-kilometer region of 30-meter sea cliffs lies between Manawainui and Ukumehame. The old road works its way along the cliffs, winding in and out of gulches, traveling over the surface of old lava flows. Unlike the old one, the new road is straight and slightly graded. Slicing through old flows, the highway cut exposes small elliptical lava tubes and beds of red and yellow ash. The road bridges the gullies and tunnels through a small section of one volcanic flow.

Suddenly the scene opens up: The shore expands into a long sandy beach, and the mountains move far back from the sea. After the constricted region of cliffs, Ukumehame is like a foothold on a mountain, and the first of the four big leeward canyons.

UKUMEHAME

Ukumehame, 'paid mehame wood', is a wide valley with a central ridge. A dike in the canyon's north wall resembles the

Ukumehame Canyon. *Inset:* An exposed lava dike on the west wall of the canyon resembles the wings of a gull.

wings of a gliding bird. The central ridge, triangular in form, sweeps gracefully upward from the alluvial fan like a wave, pauses, then rises steeply to join the high ridge at 850 meters. A well in the valley yields slightly warm water. Evidently a body of hot rock still exists underground even though the volcano itself is dead.

Also in the north wall of the canyon is the volcanic dome Puʻu Kōʻai. Squeezing out of a vent 8 meters wide like toothpaste out of a tube, the viscous lava upended older lava beds, fractured the rock, and built the 788-meter volcanic dome at the edge of the valley. Erosion has exposed the flat walls of the dike a little farther into the valley. Streams cut V-shaped gullies into the dike, shaping it into gull wings. Farther up the valley are prominent dikes running from valley floor to summit.

Rough red rocks form the walls of a *heiau* at the mouth of the canyon. Walled on two sides, it opens onto a terrace that fronts the sea. As with most *heiau* sites, the view is a commanding one, and in the quietness our imagination wanders back in time to the days when the oracle tower stood over fierce images and ritual activity enlivened this sacred place.

The 5-kilometer Ukumehame coastal plain narrows at an old sea cliff, now standing 150 meters from the shore. Then it swells into the classic alluvial fan of Olowalu ('many hills') with its conspicuous late cinder cone.

OLOWALU

Olowalu Valley is a big V-shaped cut in the mountains, its alluvial fan brilliant green with sugarcane. This region sometimes receives only 5 centimeters of rain a year; irrigation ditches form concentric arcs of circles on the fan, centering at the mouth of the valley. The soil is fertile, but it is mixed with big boulders, some of them a meter or two in diameter.

Olowalu Valley with its prominent alluvial fan. Arrow points to Kīlea cinder cone. *Inset:* Close view of Kīlea cone.

Petroglyphs at Olowalu.

Although rectangular piles of boulders dotting the fan resemble *heiau* platforms, they are rocks piled up by plantation workers during field clearing.

Olowalu Stream carved a huge valley and built a classically beautiful alluvial fan. Then volcanic activity of the late Lahaina Volcanic Series shattered the fan to build Kīlea cinder cone. Rounded in form, Kīlea looks like a yellow haystack in a canefield. The flat, slabby side of it facing the stream is covered with petroglyphs.

Petroglyphs are ancient Hawaiian carvings in stone. Two or three centuries ago, travelers "signed in" on the rock with a picture. Using stone implements, they pecked, chiseled, and carved pictures of animals, boats, lateen sails; a canoe paddler, fisherman, diver; a father with children—warm humanity of old expressed in cold rock.

A trail once led past Kīlea cone and up through Olowalu Canyon, across the divide, and down into ʻīao Valley. It was this trail that Kalanikupule used in escaping from Kamehameha after his defeat in ʻīao Valley. Landslides in the past 200 years have all but obliterated it.

Coral and shells are stuck to the rock in Olowalu Canyon 70 meters above the present level of the sea. Here, then, is evidence of a higher stand of the sea that occurred during a warming period between glacial advances. Should all the ice on Earth be melted, the sea would surround Kīlea cone, leaving it as a small island 5 meters high and lying a kilometer off shore.

LAUNIUPOKO

Beyond the smooth Olowalu fan, the mountains again come down to the sea. The next 4 kilometers between here and Launiupoko, 'short coconut leaf', is the base of a triangular sector of eroded mountain. The apex of the triangle is an 890-meter ridge between the heads of the two big valleys, 5.2 kilometers from the sea. Launiupoko and Olowalu canyons are deeply cut, the walls are high, and the sector has the appearance of a steep mountain.

Stuck on the slopes of that triangular sector and just before Launiupoko are two white volcanic domes, Puʻu Māhanalua-

The trachyte domes of Puʻu Hipa *(right)* and Puʻu Māhanaluanui *(center)*. Concentric growth rings are visible on Puʻu Hipa.

Launiupoko Canyon as seen from Launiupoko Park. The two low white cones at right are Puʻu Hipa and Puʻu Māhanaluanui.

nui, 247 meters high, and Puʻu Hipa, 305 meters. Both are eruptions of the Honolua Volcanic Series. Lava, squeezing from the Puʻu Hipa vent, formed concentric growth rings as well as ridges of pasty lava. A 10-meter cross section of Puʻu Māhanaluanui is exposed in the road cut. Marine fossil beds lie around the base of the hill, indicating that Puʻu Māhanaluanui existed long before the sea stood 80 meters higher that it does today. The dome is made of **trachyte,** a light-colored rock, poor in magnesium and iron, but rich in potassium and sodium. Part of the dome has been mined for use in making cement.

Deep Launiupoko Canyon has steep walls and, when viewed from Launiupoko Park, appears to be a V-shaped valley without a head. Rainfall is scant, soil is thin and scarcely covers the bedrock, but tiny red and yellow flowers grow on the brown hills, and rocks take on the polish of a varnished surface. The V-shaped gullies behind Puʻu Hipa and to the left of Launiupoko Canyon form chevron shapes as clouds cast shadows in them. Pūehuehu Nui Gulch, between the two canyons, is forming an amphitheater head.

KAUAʻULA

The last of the four big valleys of leeward West Maui is straight, short, amphitheater-headed Kauaʻula. It is backed up against ʻĪao Valley and separated from it by a 1600-meter ridge; their streams flow in opposite directions. Kauaʻula

Stream flows almost directly west; ʻĪao, east. Kauaʻula is typical of the box-headed canyons that develop on the short, steep slopes of West Maui. An unnamed gulch on the north joins Kauaʻula Stream at its mouth, the tributary having been diverted by the building of the volcanic dome, Paʻupaʻu.

Paʻupaʻu ('drudgery', because servants were weary of bringing water to bathe the chief's child) is also known as Mount Ball. Conspicuously different from the sawtooth gulches around it, Paʻupaʻu is an obvious late addition. Double humped when we view it from Launiupoko, Paʻupaʻu is singly rounded as we see it from Lahaina, its top rising to 685 meters. Trachyte flowed from the cone and covered the dark basalts with white rock.

LAHAINA

The Lahaina region is rich in history. Tradition tells of powerful chiefs living here in ancient times. Lahaina became the capital of Hawaiʻi in 1820; it remained the seat of government for 34 years until Honolulu took that role.

American missionaries settled in Lahaina in 1825. In 1831, they established Lahainaluna School, the oldest school west of the Mississippi. Chief Hoapili, in that same year, built his fort of coral rocks at the water's edge in Lahaina.

In the mid 1840s, more than 400 whaling ships were calling every year at Lahaina—three times as many as at Honolulu. Lahaina was then a popular whaling port, for it was an open roadstead and no pilot was needed. Later in the nineteenth century, crews of immigrant workers carved a 30-kilometer irrigation ditch in the rocky hills above Lahaina to water the sugarcane planted on the alluvial fans.

The feeling of human community at Lahaina, both past and present, is enhanced by the community of volcanic islands rising from the neighboring sea: Kahoʻolawe, Lānaʻi, and Molokaʻi.

Three *heiau* were built at Lahaina: Wailehua, Halekumu-

Creation of Puʻunoa Point. Flows from the cinder cone Puʻu Laina
(left center) diverted the courses of Kahoma and Kanahā streams
and reached the ocean in the area presently occupied by the town of
Lahaina. Paʻupaʻu is the light-colored mountain at right. Erosion
below Paʻupaʻu has since cut many gulches.

kalani, and Halulukoakoa. A resident of Lahaina, D. Kahau-
lilio, wrote of them in 1906*:

> Of the three, only the Wailehua heiau still remains. I was caretaker for
> a period of 50 years. Inserted within my house yard is the place where
> men were laid on the altar, that is to say, men selected for death were of-
> fered. On the nights of Kane and Lono, my family and I always hear the
> sound of the drum, just like the drums of the Japanese theater. This
> heiau, in the days of Lono, was filled with chiefs who looked about,
> bathed in the surf, and plunged into the water for the fragrant Lipoa of
> Wailehua, which is still found there.

*His words are quoted in W. M. Walker's manuscript "Archaeology of
Maui."

52

Two kilometers north of Lahaina is Pu'u Laina, at an elevation of 200 meters, a cinder cone of the Lahaina Volcanic Series. The eruption broke through the alluvial fans of Kanahā and Kahoma streams, forcing them southward. Kahoma has since cut a 60-meter gulch in the weak alluvium. The crater of Laina has most recently been used as an irrigation reservoir. Long ago, it was the home of the volcano goddess Pele. Peripatetic Pele once lived on Kaua'i, then at various places on O'ahu and Moloka'i. Moving down the island chain to be where the action was, she occupied the cone of Laina just before she moved on to Haleakalā. Now she lives at Kīlauea volcano on the island of Hawai'i.

A flow from Pu'u Laina poured into the sea to build Pu'u-noa Point at Māla. The flow is rich in brownish green crystals of olivine. Beyond the point and extending 250 meters out into the water is Māla wharf. Narrow-gauge railroad tracks run out to the end of the wharf. But the sea is smashing the abandoned structure, exposing steel reinforcing bars that are staining the concrete a rusty red.

Māla wharf was built in 1922 to take care of vessels that could not dock at Lahaina. Large ships had to anchor offshore and transfer freight and passengers ashore in small boats. But as soon as Māla wharf was completed, it was found to be unusable, for powerful currents made the landing too dangerous. Vessels still had to stand off a short distance from the wharf and swing their loads over and onto railroad cars. The wharf is condemned and the sea is gradually claiming it.

The view of leeward West Maui from the end of Māla wharf is excellent, just as the view of the eastern side of West Maui from Kahului breakwater is also excellent. The two structures are almost on an east-west line—a line that also runs through Pa'upa'u Hill and the summit, Pu'u Kukui. The green apron of sugarcane runs upward to the base of Pa'upa'u Hill. Kahoma and Kanahā valleys are prominent. South of Pa'upa'u Hill is Kaua'ula Canyon, then the enlarged valley head of Pūehuehu Nui, and finally Launiupoko.

To Maui's shores have come a rich blend of diverse cultures: Polynesian, European, American, Asian.

Next to Māla wharf is the Lahaina Hongwanji Jodo Mission. A large bronze Buddha, the largest outside Japan, sits in serene contemplation, eyes half-closed, fingers in a closed curve. The statue faces the setting sun, which sinks behind Lāna'i at certain times of the year. A bell in the courtyard of the mission emits a resonant sound when the ramrod clapper, suspended on chains, is pushed against it.

BEYOND LAHAINA

Northward, the agricultural crop changes from sugar to pineapple. Here is a seeming paradox—sugar, which needs more water than pineapple, growing in a drier region. Its greenness, though, is an indication not of rainfall but of irrigation. A ton of water is required to make a pound of sugar; but pineapple depends on rainfall.

The coastline indents slightly for the next 4 kilometers, then swells gently at Hanaka'ō'ō Point. Just beyond the point is Pu'u Keka'a, the last of the four cones of the Lahaina Volcanic Series. Towering high above a long sandy beach, Pu'u Keka'a was a sacred place for the Hawaiians; they constructed a *heiau* on its 26-meter summit.

Keka'a erupted a short distance offshore, sending cinder and spatter into the air to build the cone. Lava from the eruption flowed out onto the surrounding deeply weathered lavas of an earlier age. So recently did Keka'a erupt that it was not nipped at the 8-meter level by the Waimanalo stand of the sea 125,000 years ago.

Evidence of lower sea levels is also here at Keka'a. Fifty-five meters below the surface is an old bench cut into the rock by waves that washed the shore during a cold glacial period.

The cinder cone which formed Pu'u Keka'a, now called Keka'a Point or Black Rock, at Kā'anapali. Hanaka'ō'ō Point is at extreme lower right.

"Black coral" that grows on that bench is not actually a coral but is closely related to it.

A sandy beach curves around Honokōwai Point, the coast indents, and we come to the first of the Bays of Pi'ilani.

Wrapping around the northern end of West Maui are six bays, famous in song as those of Chief Pi'ilani. He ruled the bays and the three islands that he could see from them: Lāna'i, Moloka'i, and Kaho'olawe. His command of the bays is recognized in the name of the road around West Maui—Hono-a-pi'ilani, 'the bays acquired by Chief Pi'ilani'.

The bays are:

Honokōwai, a slight waver in the shoreline
Honokeana, a cavelike recess
Honokahua, a sandy, open beach
Honolua, a circular bay rimmed with high sea cliffs
Honokōhau, a wide-open V with an old red cinder cone
Hononana, sea-cliffed and catching the trade winds

Honokōwai, 'the bay drawing water', is a long, slight curve of calcareous beach sandstone—hard as rock because of the chemical cementing of its grains. Its layered surface extends into the sea to the depth of a meter, and the waves tumble rocks and pebbles on its surface. Just why beach rock forms is not clear. It may be forming at the present time, but this is hard to tell: Such beach rock probably formed in earlier times under different conditions. Soft sand, lying behind the sweep of this natural concrete barrier, is bedded at about the same angle as the beach rock.

The second of Pi'ilani's bays, Honokeana, 'the cave bay', is a deep, boxy recess in the shoreline. The sea erodes a bed of red ash on the north side of the bay, coloring the waters a murky red. The bays of Pi'ilani are of great variety and form, and these first two, Honokōwai and Honokeana, are the extremes. Honokōwai is such a slight waver in the coastline that it is hardly a bay at all; but Honokeana is so small, such a deep indentation, that it is more cove than bay. Honokeana's headlands, under the relentless attack of the sea, will gradually wear back; its material to be sorted and deposited in the embayment. Eventually, the recessed shoreline will straighten; then it may look like the Honokōwai we see today.

Just around the headland from Honokeana is the open bay of Nāpili, 'the joining' or '*pili* grass'. The half-kilometer beach, straight and sandy, has flaring ends that make it look longer than it really is. Kapalua Ridge swings far out into the sea in a gentle curve to the north. Inland, the ridge is bordered with Norfolk Island pines. A latitude line 21 degrees north of the equator runs through Nāpili Bay.

The curve of Kapalua Ridge gives Fleming Beach a somewhat closed, semicircular shape. Long waves wash far up on the shore, turning the sand a beige color at the swash marks. Running back into the sea, the water gains momentum and dissipates energy in a turbulent line of foam. All this—beach, waves, palms, colors—blend into a magnificent scene that fits the wistful concept of a tropical paradise.

Nāmalu Bay, 'the protections', a bay without sand, is wedged between the northern point of Fleming Bay and Hāwea Point. Standing on this northwest point of Maui is a lighthouse. Here the long, sandy stretches of beach on the leeward coast of deposition give way to the windward coast of

The West Maui coastline swings northward from Kaʻelekiʻi Point *(extreme right)*. To the left of the point is Nāpili Bay, followed by Kapalua (Fleming Beach), then Nāmalu Bay and Hāwea Point. In the distance is Līpoa Point. Makāluapuna Point is the slender finger of white between Hāwea and Līpoa points.

active abrasion where old lava flows jut clawlike and prehensile into the sea.

The 3-kilometer stretch from Hāwea Point to Līpoa Point is a large embayment, cradling within it four smaller bays. The two long, straight, north-facing bays of Oneloa and Honokahua were once joined. But during the time of Honolua eruptions, vents 8 kilometers up the mountain poured out trachyte lava that ran into the sea, forming Makāluapuna Point and creating the two bays.

Salt spray has etched the whitish gray trachyte of the peninsula of Makāluapuna into sharply pointed forms like dragon's teeth. The headland is a flat, windy park of great varie-

Trachyte flowing into the sea, shown here in black, bisected the long beach at center, creating Makāluapuna Point and two bays: Honokahua and Oneloa. The point beyond the black lava flow is Hāwea, and the dome in the distance is the shield of Lāna'i Island. Lower left is Honolua Bay, and just above it, Makuleia Bay. *Inset:* Makāluapuna Point as it appears today.

"Dragon's Teeth" at Makāluapuna Point.

ty. Short-cropped grass and *naupaka* carpet the thin soil. Big boulders weathering brown and butterflies flitting among tiny blue flowers make Makāluapuna Point a natural garden.

The village of Honokahua lies behind Makāluapuna Point. It is a typical old plantation village; its cottages have rust-red roofs and white window frames; grassy yards are bordered with ti plants, banana trees, and plumeria. Such earthy plantation villages are a part of the experience of thousands of people of Hawai'i.

Above the village, a double line of Norfolk Island pines set 8 meters apart borders the road from Nāpili to the old plantation house at Pineapple Hill. This is a place of pineapple fields and rolling hills, of broad swinging contours, of slopes that run upward to old cones and on into the clouds.

Across Pailolo Channel from Honokahua is the island of Moloka'i, 15 kilometers to the northwest. Lāna'i, to the southwest seems to be floating on the ocean like a surfaced whale.

A wall-like mass of trachyte 30 meters high rises boldly from the water at Līpoa Point. Tucked behind the point is the fourth of Pi'ilani's bays: Honolua, 'two harbors'. The bay is rounded, and a twin of Makuleia Bay. Together they do indeed form two harbors. A wall of Honolua lava protects Honolua from the northeast trades. Honolua Stream works its way to the sea through a pebble, cobble, and boulder beach. Coral heads show through the clear water, and schools of *akule* move like a dark cloud in the green water above the sandy bottom.

We may pause here to watch a fishing operation. From a *kilohana*, an observation post high above the bay, a man signals to the fishermen below. Cautiously, silently, they push a boat loaded with nets across the bay. Tension mounts. Will they get the net spread before the fish sense the trap? Suddenly the dark cloud of fish disappears. Instantly. Much too fast for anyone to see. Silver flashes in the closing ring indicate a successful catch, and the net is gathered in.

In ancient times, nets were made of scraped and rolled bark of the *olonā* bush; *kukui* wood floats supported them in the water. Modern nets, though, are of nylon, and empty plastic bleach bottles serve as floats. The paddlers now wear rubber swim fins as they push their fiberglass boat across Honolua Bay.

Here in Honolua Bay, in April, 1976, the Polynesian Voyaging Society's double-hulled canoe *Hokule'a* spent a few days just before its historic trip to Tahiti. The purpose of the voyage was to test the question: "Is two-way voyaging between Hawai'i and Tahiti possible, employing ancient methods of navigation and a type of sailing craft that was used by the early Polynesian explorers and settlers?"

The success of the experiment answered the question affirmatively, ending speculation. The ancient craft was capable of sailing close enough to the wind to make the trip, and the navigational system was accurate enough to allow the Polynesians to sail thousands of kilometers over the open ocean to intended landfalls.

A late trachyte flow pushed into the sea to form Līpoa Point, and built the windward side of Honolua Bay *(extreme right)*.

A windward course took the *Hokule'a* around the east side of Maui and Hawai'i, the course chosen to afford the vessel sufficient "easting" for the Tahiti landfall. A year later, an experimental voyage through Kealaikahiki Channel and a few hundred kilometers out to sea confirmed that this channel, as its name suggests, is truly "the path to Tahiti."

Līpoa Point, 'seaweeds', is a broad, flat region of Honolua lavas planted to pineapple. It's a good hike around the point. Winds are brisk. Ironwood trees are windbreaks that lend a fragrance to the air as their long soft needles cushion the ground. Waves washing into coves in Līpoa Point have hollowed out caves.

Look back into Honolua Bay from Līpoa Point and you can clearly see the white trachyte of the Honolua Volcanic Series veneering the black basalts of the old Wailuku shield. Trachyte forms late in the life of a Hawaiian volcano. In the cooling magma chamber, the denser minerals, rich in magnesium, calcium, and iron, settle out and are carried downward. Minerals of less density—those containing sodium and potassium—remain at the top. Renewed volcanic activity may form fissures that tap any part of the chamber. If the activity happens to tap the top of the pot, trachyte may surface, as it has done in the Honolua volcanics.

Heakalani *heiau* stands on a hill above a gulch at the north end of Maui. Measuring 60 meters in length, Heakalani is one of the eleven largest *heiau* on the island. Its importance is indicated by a *luakini*, a place of human sacrifice. The surface of the *heiau* is composed of stream-worn cobbles and coral pebbles, and a trail paved with stones runs down to the beach.

The fifth of Pi'ilani's bays, Honokōhau, 'the bay drawing dew', lies at the mouth of West Maui's longest stream. From its head, an amphitheater just below Pu'u Kukui, Honokōhau Stream flows 15 kilometers straight north in a troughlike channel. The surface of the land slopes into the sea to meet Moloka'i 80 meters beneath the surface of Pailolo Channel.

The waterfall at the head of Honokōhau drops 520 meters. This waterfall, almost inaccessible, may be the sixth tallest in the world, second in the nation after Yosemite in California.

The road into Honokōhau Valley runs along the top of a 45-meter sea cliff. A bright yellow layer of volcanic ash is at

Pu'u Ka'eo is in center foreground, and to the right is Honokōhau Canyon. Mount 'Eke is obscured by clouds in this photo.

the water's edge. The road loses elevation as it goes upstream and the ash bed gains, so that the two intersect and the 3-meter layer is well exposed. This prominent marker bed of brilliant yellow pumice has draped itself over preexisting rocks of the Wailuku Volcanic Series and, in turn, has been covered by Honolua lavas.

The very old cinder cone at the east side of Honokōhau is the 108-meter Puʻu Kaʻeo, a cone of the Wailuku Volcanic Series. Much of it has been eroded by the sea. Exposures of ash and pumice—brilliant red and lavender in color—suggest a fiery origin for the ancient cone.

Between Honokōhau and Nākālele Point is a large, slightly sloping triangular region extending 3 kilometers inland, Kulaokalālāloa. The sector is old and represents the original completed surface of the West Maui shield. Now an agricultural area, the seaward edge of Kulaokalālāloa is rimmed with a windbreak of ironwood trees. A long, dusty plume of red dirt trails off toward the leeward as pineapple machinery works the soil. A line of sea stacks extends a kilometer out from Kanounou Point near Puʻu Kaʻeo. On a clear day, Mount ʻEke shows its stepped series of lava flows.

Nākālele Point pierces the ocean in a sheaf of stacked aa flows that stands like a brick wall. Little if any soil covers the point. The bare rock is a sharp contrast to the smooth cultivated fields of Kulaokalālāloa with its fringe of windbreak.

A blowhole on the east side of Nākālele Point bursts into whiteness as the air is trapped and compressed in a sea cave by incoming waves, then released through a hole in the chamber's roof with a hollow whooshing sound. Beyond the bloom of the blowhole is Keawalua, a flat-floored gulch with steep sides and a stone corral—Anakaluahine Gulch, ʻcave of the old lady', near the sea.

The region seems desolate. The sun is warm, and there's always a breeze, sometimes so strong that you can lean into it (*nākālele* means 'leaning'). Fanciful wind forms carved from ancient lava flows suggest the mesa country of the southwest American desert.

High above the breaking waves, far from the shore, are cliffs that resemble sea cliffs; but the sea has never been this high since West Maui was built. Stained reddish and beige, these wavy cliffs were carved by the wind in the residual soil of Wailuku basalt.

Vegetation struggles to exist against the wind. Headward migration of gullies in the rugged area exposes red soil. Erosion scars are not healed over, for the wind blows away new soil particles as fast as they can form. Brown boulders roll down the embankments and rest on the gully wash. Eddying winds blow the sand away from the bases of the boulders to leave them standing like pedestals.

Gulches are dry. But Hawaiians making the trip from Waiheʻe to Honokōhau knew how to find water. Guided by knowledge and perhaps intuition, the traveler would select a likely place. By removing the rocks and then digging in the gravel, he'd uncover the source of water. When finished, he'd stack the rocks back in place, carefully, insuring the purity of water for future needs.

An old Hawaiian trail, 2 meters wide and edged with curbstones, follows a straight line over the hills and through the valleys. No attempt at contouring was made to decrease the

Wind-carved terraces near Nākālele Point. Kahakuloa Head is in the distance.

Remnants of the old Hono-a-Pi'ilani trail.

grade. What remains of the trail is clear and firm, but a combination of gullying and bulldozing has obliterated most of it.

Long swells enter the shallow, sandy Po'elua Beach. Between dry gulches stand tall cliffs. By extending the line of lava slopes, you can estimate the amount of land worn back by the sea. From here to Kahakuloa, the cliffs rise gradually to greater heights.

Near Hononana is a 3-meter boulder beside the road. When struck with a rock, Pōhakukani, 'the bell stone', gives forth a sonorous sound remotely resembling the ringing of a massive bell—solid, but without metallic quality. Curious travelers searching for the responsive note have worn shallow pits in Pōhakukani.

KAHAKULOA

The windblown, open region of West Maui, where the feeling is rural, expansive, and free, ends at Kahakuloa in a 6.5-kilometer constricted region of high sea cliffs.

Kahakuloa Head and Puʻu Kāhuliʻanapa. The shield of Haleakalā rises in the background.

Kahakuloa is West Maui's most prominent seacoast feature, a volcanic dome high as a 60-story building. Set apart from the rest of the volcanic slopes, Kahakuloa, 'the tall lord', resembles a majestic royal figure in a feather cloak. If the 194-meter Kahakuloa is the tall lord, then Puʻu Kāhuliʻanapa, a 165-meter volcanic dome across the swale to the south, might well be his attendant. Together they are an imposing pair.

Kahakuloa is the steeper of the two. When seen from the road a short distance away, its sides seem almost vertical and much too steep to climb. But its apparent steepness is partly an illusion. The ascent can be made, using hands as well as feet, a good head, and a measure of courage.

Halfway up on the seaward side is an area of fretwork weathering. Frets stand as partitions between pits in the rock 3 or 4 centimeters in diameter and twice as deep. The etching of the rock is a combination of salt spray and the action of organisms—organisms that hold tightly to the rock and in growing dissolve it bit by bit.

Thick, yielding grass, wind-pruned shrubs, and scattered tiny flowers lend an alpine appearance to Kahakuloa. Snail shells whiten in the sun. Lantana, a beautiful flower on a

Fretwork weathering at Kahakuloa.

thorny bush, spreads uncontrolled. The thorns of this head-high bush are a far greater deterrent to the hiker than the steepness of the slope.

Puʻu Koaʻe, 'tropic bird hill', is the summit of Kahakuloa. The white-tailed tropic bird with a wingspread of a meter flies along the coast. The wedge-tailed shearwater skims the surface of the sea sending out a call that sounds like a whistling tea kettle. Greatest of all the birds here is the black, slender frigate bird with its wingspread of 2 meters, soaring high on thermal updrafts.

It is windy on top of Puʻu Koaʻe, but a few paces to the leeward there's hardly a breeze at all, and you feel the warmth of the sun. The trade winds are strong at low altitudes, fresh and invigorating after having traveled over thousands of kilometers of open ocean. The warm, moist air blows past Kahakuloa, rises, and condenses against the West Maui Mountains in a thick cloud—so thick that the forests beneath look black. The peak within the clouds, Pu'u Kukui, receives about 988 centimeters of rain a year, making it the third wettest place in Hawai'i and ranking high among the wettest places on Earth.

Sea and sky meet at the horizon, 60 kilometers distant. The next land of any kind, the North American continent, is 80 times farther. We know what lies beyond that horizon, but the Polynesians who stood on this rock a thousand years ago must have wondered what islands lay in the direction of the prevailing winds.

The horizon appears to be at the same level as Pu'u Koa'e. From the horizon, the sea seems to slope downward to the base of Kahakuloa. Looking straight down, you can watch great boulders in the shallow water change shape as waves curl over their tops.

MOUNT 'EKE

A volcanic dome lies at each end of Kahakuloa Stream. Kahakuloa is at the mouth of the stream, Mount 'Eke at its head. 'Eke is a thick, steep-sided dome with a flat top looking like

Mount 'Eke and Honokōhau Canyon.

an upside-down cupcake. The dome rests on three flows of trachyte, each 120 meters thick.

Rainfall on the 1370-meter summit of Mount 'Eke averages 650 centimeters yearly. Such heavy rainfall in a tropical land usually means dense vegetation. But here on 'Eke, as in other high bogs in the Hawaiian Islands, vegetation is stunted. A dwarfed tree half a meter high—a natural *bonsai* —might be a hundred years old. Silversword abounds, a close relative of the silversword of Haleakalā. Gnarled *'ōhi'a lehua* is only a low-lying shrub.

The rimless summit of 'Eke is 800 meters in diameter. Spongy vegetation covers the surface, resting on black mud, a peat bog resulting from the extreme acidity of the soil. Beneath the peat is a layer of yellow limonite streaked with clay. Beneath that is the trachyte.

Water holes 2 to 6 meters in diameter at the center of the summit rest on bedrock. Deep, narrow crevices near the edge of the plateau, often bridged with vegetation, make hiking hazardous. Sinkholes, from 2 to as much as 30 meters across and 25 meters deep, were probably formed by rainwater solution. Ordinarily we do not find sinkholes in lava rock, but

Summit of Mount 'Eke. White spots are sinkholes.

in the unusual conditions on 'Eke, the trachyte acts as an impure limestone, and sinkholes do form.

The dome has withstood intense erosion. It is isolated, surrounded on three sides by precipices 300 to 800 meters high. Usually the summit of 'Eke is in the clouds.

WINDWARD WEST MAUI

Along windward West Maui, whitish gray flows of the Honolua Volcanic Series cover the darker Wailuku basalts like frosting on a cake. We would expect, here in this region of heavy rainfall, to find huge canyons comparable to those on the leeward side—Launiupoko, Olowalu, and Ukumehame —and even bigger. The answer to the seeming paradox is in the protective armor of Honolua lava, which resists erosion and covers the old basalts. So the rainy side of West Maui is less eroded than the dry leeward side.

Between Kahakuloa and Hakuhe'e, sea cliffs are 150 meters high. The cliff is very sharp at Hakuhe'e. Mōke'ehia Island is a bit of Honolua lava separated from Hakuhe'e by the action of the sea. Seen from Kahului, it appears to be the end of the mountain.

White domed Pu'u Makawana is the rounded sea cliff in the foreground; Mōke'ehia islet is off Hakuhe'e Point, and Kahakuloa Head is in the distance.

Bulky, blocky Puʻu Makawana sticks toelike into the sea. Steep-sided, it is like a round building capped with a white top. On its north side is a sea cave. The source of the flow that frosted Puʻu Makawana is Puʻu Ōlaʻi , a volcanic dome 30 meters high, 300 meters from the ocean.

Sea cliffs taper southward toward the Isthmus, and at Waiheʻe Point the coast changes abruptly. Sand dunes replace cliffs and we enter a region of deposition and coral reef. Paralleling the coast is a long sand dune that separates Waiheʻe Village from the sea. Bordering the streets of Waiheʻe are weathered wooden buildings and, in season, brilliant red poinsettia.

Waiheʻe River ('slippery water', because of the algae in it) flows in a canyon 1000 meters deep. Its head is an amphitheater just below Puʻu Kukui and adjacent to Honokōhau Stream. Eventually, the two streams will join at their heads through a process of stream piracy.

On the left is the amphitheater carved by Waiheʻe Stream. Eventually its headwaters will capture Honokōhau Stream *(center)*. Puʻu Kukui is the highest point visible; white areas on its ridges indicate peat bogs.

Millions of years ago, Waiheʻe River must have been a stream flowing in a gully down the northeast side of conical West Maui volcano. With high rainfall at high altitudes and low rainfall at low altitudes, the head of the valley enlarged rapidly. Plunge-pool action on alternating resistant and non-resistant rock layers sloping downstream, along with stream piracy at the upper parts of the drainage, formed the amphitheater. Waiheʻe River, more vigorous than Honokōhau Stream, will eventually capture the other's waters above the Honokōhau waterfall.

ʻĪao Stream leaves its canyon, flows across the alluvial plain past Halekiʻi and Pihana *heiau*, makes an S-shaped curve, and enters the ocean in the delta at Nehe Point. Sediment carried by the stream colors the ocean a brownish gray; and currents sweeping past the point swirl the material northward, depositing it along Paukūkalo Beach.

4
The Isthmus

One great curve sweeps down from the top of Haleakalā, across a central plain, and to the top of West Maui.

This central plain—the Isthmus—is the low point of the sweep, like the catenary of a cord sagging between two great stanchions.

The Isthmus is a bridge between Maui's two volcanoes. Many times in the geologic past this tenuous tie was broken as invading seas washed between the two mountains, causing them to stand as separate islands in the sea.

Haleakalā gave the Isthmus its underlying structure; the surface soils came from West Maui. Lava flowing down Haleakalā shield in ancient times ponded against the preexisting West Maui volcano, creating the flatness. Streams carrying earthy sediments out of West Maui canyons covered those old flows with alluvium hundreds of meters deep close to the slopes, tapering out to thinner layers with distance.

Streaked across the Isthmus and parallel in direction to the northeast trade winds are ridges 60 meters high that taper toward the south. Thirty thousand years ago, these hills were active, moving sand dunes. Now they're fixed in place, unmovable—turned to stone.

It happened at a time when huge ice caps covered the polar regions of Earth. Sea level had been dropping for centuries as the Earth grew colder. Water evaporating from the sea fell as snow, changed into ice, and was locked into the growing polar caps. Thousands of meters in thickness, the north polar ice sheet spread across North America as a **continental glacier**, advancing as far south as the Ohio River.

While continental glaciers moved outward from the poles of the Earth, mountain glaciers crept down the summit of Mauna Kea, 150 kilometers southeast of Haleakalā. Flowing slowly down old stream channels like a mighty viscous lava flow, the mass of ice scraped the mountain top, piling the debris into hills of rubble called moraines. Scratches, striations, and polished surfaces on bedrock also remain, attesting the power of a mountain glacier in grinding and smoothing a volcanic peak. Haleakalā may have had a perpetual field of snow but certainly no glacier; for its highest point during that period was at just about the same level at which the Mauna Kea glacier was wasting away.

The lower sea of glacial times exposed great broad stretches of sandy beach at the Isthmus. For centuries the

Cross section of sand dune as seen at a road cut near Waiheʻe. Stratified layers form on the lee side of dunes, and crossbedding indicates changes in wind direction at the time of deposition.

trade winds blew across the beaches, piling the sand up into long ridges, sorting it into fine and coarse layers. Shifting winds, working and reworking the old sand layers, have produced fascinating patterns of **crossbedding.**

Vegetation anchored the drifting dunes. Plant roots, releasing carbonic acid, changed the calcium carbonate sand into a soluble bicarbonate form. Percolating through the dune, the calcium in solution travels until it is reconverted into insoluble calcium carbonate, cementing the sand grains together. The sand dunes of the Isthmus are now **lithified,** having been turned to stone in the cementing process. But nothing is permanent. Wind and water are now wearing away at the dunes, and technology hastens the process.

Walk on the sand dunes and you'll find the surface hard. But where it has been broken, the less-consolidated sand beneath is brushed out by the wind, drifting into pockets. Tiny flowers with pale petals and yellow centers vibrate with the wind in the reworked sand. Root molds, hardened into tubelike structures, stand as pedestals where loose sand has been blown away.

The armies of two great chiefs fought here on the dunes of Wailuku at about the time of Captain Cook's visit. Even now, the bones of those ancient warriors erode out of sepulchral sites in the dunes. Kalaniʻōpuʻu, chief of Hawaiʻi, tried many times to gain control of Maui. He did succeed in taking Hāna, holding it for 20 years. But when he met the Maui chief Kahekili on the dunes of Wailuku, he was defeated and returned to Hawaiʻi. Kahekili succeeded in gaining control of all the other islands except Hawaiʻi. His kingdom began to break up, though, with his son's defeat by Kamehameha at ʻĪao Valley.

Our tour of West Maui began with the view of ʻĪao Canyon as seen from Kanahā Pond and from the Kahului breakwater. Before we explore the Isthmus, let's pause to have a closer look at the pond and the two important *heiau* of Halekiʻi and Pihana.

HALEKIʻI AND PIHANA

Standing on lithified dunes 30 meters above ʻĪao Stream, the *heiau* of Halekiʻi and Pihana may have been in use during the time of Kahekili's war. Halekiʻi and Pihana are among the eleven largest *heiau* on Maui. Most of the island's 250 *heiau* are about 30 meters long. A few are small platforms 5 meters square.

Halekiʻi, 'house of images', is a platform 100 meters long and 50 meters wide. Its condition has been altered by the hand of man as well as the scythe of time. Many *heiau* were destroyed, and idols and images all over the kingdom were toppled, upon the abolishment of the *kapu* system in 1819. Halekiʻi was restored in 1958 under the direction of Bishop Museum anthropologists. A meander of ʻĪao Stream is cutting into the base of the hill. Eventually, the sides of the lithified stone will slump; the boulders will once more return to the stream, and the sand to the sea.

A lane bordered by *koa haole* leads from Halekiʻi to Pihana Kalani, 'a step toward heaven'. Only slightly smaller than Halekiʻi, Pihana is a terraced platform 100 meters by 40 meters. Terracing extends the natural level of a hill, afford-

Detail of a wall, Pihana *heiau*.

ing a larger platform and a more impressive *heiau*. The walls of Pihana are made of water-worn boulders 50 to 75 centimeters in diameter, fitted together in a tight pattern without mortar. Now spotty white with the growth of lichen and algae, the rounded rocks resemble stacked bags of rice or cement. So smooth the facing, so straight the terraced walls of the *heiau*, that we view with great admiration the skill and industry of those who built it.

Building a *heiau* was a community effort that took a good deal of manpower. A boulder 50 centimeters in diameter weighs about three times as much as a full-grown adult. To raise the boulders 40 meters, from the stream to the platform, would require ingenuity as well as industry. A trail 2 meters wide marked with cobbles and small boulders runs down to the stream, but there is no evidence of a haulage way.

The *heiau* was paved with small, flat, stream-worn pebbles

No two *heiau* were exactly alike, but details of all were similar. Usually situated on a commanding site, they were constructed of platforms of stone, often terraced. Various grass buildings, an oracle tower, an area for sacrifices, free-standing carved wooden images, and fenced enclosures completed the temple. Wrappings of tapa on poles warned all comers of the strict *kapu*. Important *heiau* were often the repository of bones of deceased kings and chiefs.

and sand. Woven *hala* mats covered the floors of the huts, fierce wooden idols lined the edge of the *heiau*, and the lofty oracle tower completed the edifice—impressive when seen from almost any place on the Isthmus. The view from the *heiau* itself was a commanding one. From Haleki'i and Pihana you can see a large portion of West Maui from Waihe'e to Mā'alaea, and the tremendous bulk of Haleakalā from sea to summit.

KANAHĀ POND

Kanahā Pond lies at the north end of the Isthmus, separated from Kahului Bay by a **bay mouth bar**. Rich in plant life—bullrush, spike, seed-producing rush, and the rush with the three-cornered stem—Kanahā Pond supports a large population of birds. It is the home of the Hawaiian stilt, *āe'o,* and the Hawaiian coot, *'alae kea.* And it is a quiet place except for the pleasant chatter of the birds as they line up as if to gossip.

Stilt birds land gracefully, their long legs trailing out behind them; in searching the edges of the pond for food, they look as if they're walking on water. Small forms of life dash around the surface of the pond like blinking lights. Dragonflies in tandem light on the branch of a bullrush. The grass is thick and spongy around the edge of the pond. But during dry periods, hundreds of square meters of pond bottom are exposed, revealing three-toed bird tracks imprinted in mud. Wind picks up the dust from curling polygonal mudcracks, coating and dulling the leaflets of the *kiawe* trees.

Both swimmers and nonswimmers use the pond—waterfowl that sit and tilt; shorebirds that wade and bend. Migratory birds, the pintails and shovelers, come from as far away as Canada and Alaska to winter at the pond. And the wandering tattler, ruddy tern, and sanderling also make their homes here. Daily movement of the birds may take them up into the mountains, out across the Isthmus, or even to the other islands within sight.

Absent from the pond is the state bird of Hawai'i, the *nēnē*, Hawaiian goose. Medium-sized, heavily barred, this goose with a black face, head, and nape of neck is a highly specialized individual who lives in regions of rugged lava flows. Because it has spent so much time on the sparsely vegetated lava, the *nēnē*'s evolutionary specialization is a reduction of the webbing between its toes. It prefers the wet upland forest where the heavy and misty rain falls. Haleakalā Ranch has set aside an area for the *nēnē*, now becoming reestablished after near extinction. Raised at Pōhakuloa in the Saddle region of the island of Hawai'i, some *nēnē* were released on Maui. The introduction of new strains is beneficial in freshening the gene pools, preventing "stagnating blood" that may occur where inbreeding is too close.

KEĀLIA POND

To the south, at the opposite end of the Isthmus, lies another pond—Keālia, 'salt encrustation'. A long strip of land, a bay

Keālia Pond and bay mouth bar at the southern end of the Isthmus. To the right is Kīhei; in the background, the mass of Haleakalā.

mouth bar, separates it from the sea. This pond, too, is the home of many birds. Now 1.5 kilometers long and 1 kilometer wide, it will eventually be filled with sediments coming from Haleakalā.

WATER FOR THE ISTHMUS

Ancient Hawaiian wells were made by blocking crevices near the beaches with rocks and straw, damming up the fresh water seeps. Another type of well was made by excavating loose clinker and aa, lining it with big boulders to keep it from caving in, then letting it fill with fresh groundwater.

With today's technology and a better understanding of the underground water table, modern wells are thousands of times more efficient than the old Hawaiian ones. The largest water pumping station in Hawai'i is located in the middle of Maui's Isthmus. This "Maui-type well" has a long shaft that penetrates the earth at a 30-degree angle, terminating in an underground pump room overlying a sump. A horizontal sa-

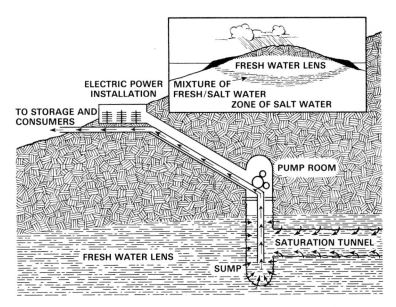

FRESH WATER LENS

ELECTRIC POWER INSTALLATION

MIXTURE OF FRESH/SALT WATER

ZONE OF SALT WATER

TO STORAGE AND CONSUMERS

PUMP ROOM

SATURATION TUNNEL

FRESH WATER LENS

SUMP

A "Maui type well" consists of a shaft sunk at a 30-degree angle, a pump room atop a vertical sump, and a horizontal saturation tunnel. The shaft may be 3 to 5 meters in diameter and accommodates a small cable car and the uptake water pipe. The saturation tunnel, which may be as much as 300 meters in length, skims water from the top of the fresh water lens. *Inset:* Diagram of an island's water supply. The lens is not an underground cavity filled with water, but rather the volume of water contained in the island's porous rock base.

turation tunnel leads off from the sump to collect water from the underground supply. The well is located 3 kilometers out from Waikapū Valley, but the water it draws comes from Haleakalā. The well has a pumping capacity of 150,000 tons of water a day. Every day, about 9 million tons of water fall on Haleakalā. Four percent runs off in streams, and much of that is diverted to irrigation.

Most of the water that falls on Haleakalā sinks into the ground. Moving downward through openings in the rock, it encounters salt water that saturates the whole base of the island. Fresh water floats on top of that denser salt water in a lens-shaped body that extends under the Isthmus. The well taps that lens of fresh water. Much, or most, of the fresh water beneath the Isthmus is recharged by irrigation water brought to the Isthmus via the East Maui ditches, rather than by direct flow from the slopes of Haleakalā.

Although the Isthmus is dry, irrigation makes it arable. With the development of plantation agriculture, villages— clusters of houses for the workers—were established at locations convenient to the fields. Just in the names of these little villages we can get an idea of the introduced cultural influences which have shaped modern Maui—everything from colorful, personal, descriptive names to banal numbering.

Yung Hee Village	Cod Fish Village
Village 5	Camp K-3
Alabama Village	Orpheum Village
Spanish B Village	Hawaiian-Spanish Village
Sam Sing Village	Pu'unēnē
Village 4	Kaheka Village
Upper Village 3	Store Village
Middle Village 3	School Village
Japanese Village 1	Airport Village
Hawaiian Village	McGerrow Village
Russian Village	Nashiwa Village
Spreckelsville	Ah Fong Village

5
Haleakalā

Haleakalā, the East Maui mountain, is triangular in shape, each side about 40 kilometers long. The points of the triangle mark its three major rifts—north, east, and southwest. The whole shield is known today as Haleakalā. But to the Hawaiians of old, Haleakalā was only one small peak near the top of the mountain.

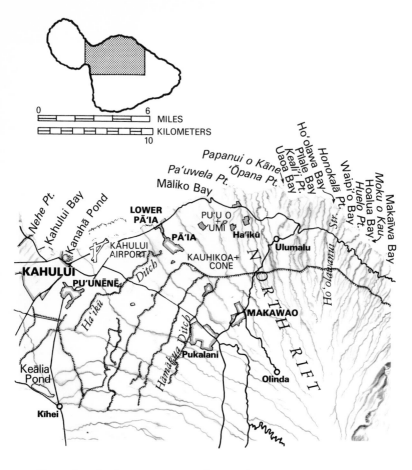

Captain Cook, writing on November 26, 1778, was the first European to describe Haleakalā:

An elevated hill appeared in the country, whose summit rose above the clouds. The land, from this hill, fell in a gradual slope, terminating in a steep, rocky coast; the sea breaking against it in a most dreadful surf.

Haleakalā rises 8 kilometers from its base on the ocean floor. We see the peak of the shield—perhaps 7 percent of its mass—rising 3 kilometers above the sea. Fifteen times as much as we can see lies beneath the waves.

Younger by far than the volcano that forms the West Maui

Mountains, Haleakalā's slopes are smoother and flatter. Last active about two centuries ago along its southwest rift, Haleakalā's most recent flows are still black and fresh looking, little altered in this region of slight rainfall.

The windward side of the mountain is cool and wet; the leeward side, hot and dry. The side facing the Isthmus, though, is a varying blend of the two.

WINDWARD HALEAKALĀ

Dense forests cover the rainy side of Haleakalā from Pa'uwela Point to Hāna and around to Kīpahulu. Vegetation is so thick and lush in spots, and shade is so deep, that only fern and basketgrass grow on the forest floor. Guava, dominating at lower elevations, gives way to *koa* at about 750 meters.

Moisture carried by the trade winds condenses to form clouds that wrap around the peak. Immediately beneath that ring of clouds the climate is cool, foggy, and moist—like the cool marine climate of temperate zones.

Pā'ia, 'noisy', is like a frontier town nestled in farmlands, a supplier of goods and social life to the community. Its sugar mill, of classic form, is one of the few in Hawai'i to have installed a "diffuser"—actually a washer or leacher. After the cane has been shredded, the diffuser washes the sugar out of the fiber. The fiber residue (bagasse) is then burned in the factory boilers to produce steam and power. Canefields surround the town and run down to the sea at Lower Pā'ia. Waves break on the reef a kilometer offshore, and sandy beaches fill the cusps between the rocky headlands.

A fringing reef, not of large extent, starts at Lower Pā'ia and widens into Kahului Bay. For years this was a place of marine deposition, and a beautiful beach of limey sand formed the shore. But conditions changed; the beach is washing away. Some local residents think that the disappearance is due to the building of the breakwater in the early 1900s.

Modern land usage practice may also have had an effect. Sand and silt, carried by streams down to the sea, is swept

West Maui and the Isthmus as seen from near Spreckelsville. Kanahā
Pond is at center of photo, with Kahului, Wailuku, and ʻĪao Valley
beyond. In the foreground, waves break over the fringing reef.

around Paʻuwela Point by powerful currents and deposited
in the bay. Storms also stir up the bay and keep it muddy for
days or weeks. Organisms with low tolerance to such conditions die, and the reef cannot grow.

Pāʻia lies on the north rift of Haleakalā. Beyond the village, and for the next several kilometers along the coast, is a
group of cinder cones running from sea to summit. Covered
with trees, they are dark green hills dotting fields of light
green sugarcane. Formed during the middle period in the
building of Haleakalā, these old cones, although similar to
those of the Hāna Volcanic Series, are of the Kula series, indicating that the north rift did not open during the most recent
period of volcanic activity on Maui.

A deep indentation in the coastline 5 kilometers beyond
Lower Pāʻia is boxy Māliko Bay ('budding'). The road runs
along a 30-meter cliff, then turns upstream for a suitable
place to cross the bay.

Māliko Bay. The relatively level fields on either side of the bay have been further flattened by present-day cultivation. Note sea stacks.

Here in the banks of Māliko Stream is one of the few places on the island where lavas of the oldest volcanic series (Honomanū) are exposed. Fifteen hundred meters farther upstream is bulky Puʻu o ʻUmi cone, its summit at 193 meters.

Māliko Gulch deepens into a canyon at Kauhikoa cone near the town of Makawao. A century ago, the "fearsome gorge of Māliko" was an almost impossible barrier for Samuel T. Alexander and Henry P. Baldwin in their building of Maui's first large irrigation system, the Hāmākua Ditch. A time of great agricultural effort, this period saw widespread construction of irrigation systems to increase the productivity of this and other regions of Maui.

Borrowing $80,000 from Castle and Cooke, Alexander and Baldwin began their Hāmākua system to bring water to the Isthmus. Completing 27 kilometers of ditch, tunnel, and flume was a monumental task in itself. But added to that already existing difficulty was the fact that they had to complete it in a limited time or lose it all to the "sugar king," Claus Spreckels. The task of high venture demanded the utmost effort from those who helped complete it.

Work went well for Alexander and Baldwin on the Hāmākua Ditch until they reached Māliko Gulch. To cross such a canyon, men were lowered on ropes to prepare the channel and lay the pipes for the siphon. But the straight walls and depth of Māliko were so terrifying a sight that none of the men would volunteer for the job.

Finally Baldwin himself became the hero of the project. Handicapped by the loss of an arm in a sugar mill accident, he used his one arm to hold onto a rope. Slowing his 120-meter descent into the deep gorge with his knees, he reached the bottom safely. A cheer broke out for his accomplishment, and the other men soon followed his lead. The Hāmākua ditch was completed just in the nick of time in a melodramatic finish. The water was turned on; a day later, it reached the parched Haʻikū plantation, only one day before the deadline set in the royal grant. Soon a quarter of a million tons of water was flowing each day through the irrigation system.

Ten kilometers up Māliko Gulch, at an elevation of 480 meters, is Makawao, 'forest beginning'. A cattle-raising town surrounded by sugarcane fields, Makawao uniquely combines Western and Oriental traditions. Buildings along the main street have the weathered look of old cattle towns of the American West, yet the names on some of the tall false fronts are Oriental, and written in *kanji* characters. Hawaiian cowboys saunter along the street, country-western music drifts out of cafés, and you can buy "shave ice" in the general store. The richness, diversity, and blend of various ways of life are seen nowhere better than at Makawao.

Above Makawao, Māliko Gulch is only a slight indentation on the surface of the land, and the stream itself but a trickle. Paperbark and eucalyptus trees border the winding road toward Olinda among the Kula cinder cones of the north rift. Meadows and rolling grasslands alternate with clumps of trees in woodlots. Olinda, at 1100 meters, is above the level of maximum rainfall.

Returning down Māliko Gulch from rain forest to seacoast, we pass a large reservoir at Haʻikū and find a lighthouse standing on flat Paʻuwela Point. The 4-kilometer stretch between Paʻuwela and ʻŌpana Points, slightly embayed, runs almost directly eastward. Paralleling the coast 3 kilometers inland is the road. Another 2 kilometers inland from the road is Ulumalu village with its prominent 350-meter cone, Kapua'i o Kamehameha. Behind the wooden fences that line the road are redwood water tanks set on stilts. In deep road cuts, 10 meters of reddish brown soil is exposed.

Papanui o Kāne is an islet—a sea stack—at 'Ōpana Point. It marks the entrance to the broad, open, U-shaped Uaoa Bay, the largest along this coast. Keali'i Point, 2 kilometers from 'Ōpana Point, forms the far side of the bay. Pīlale Bay follows, and the coastline becomes more rugged as it faces directly into the northeast trade winds. Sea cliffs here are 30 meters high and will continue to increase in height for the next 20 kilometers.

Ho'olawa Bay is a deep, boxy recess in the shore into which Ho'olawa Stream pours, fed by the upland twin streams Ho'olawanui and Ho'olawali'ili'i. A hike 1000 meters upstream from the road brings you to the confluence of the two and to a twin waterfall, where water drops over a dark lava ledge.

The base of Honokalā Point knifes into the sea; Waipi'o Bay is tucked in behind the point. Broad and open, Waipi'o is much like Uaoa Bay but not quite so large. Here the coast takes a definite swing to the southeast, swelling into Huelo Point—a broad, flat area 1000 meters across and 100 meters high.

The old Kaulanapueo Church, built in 1853, stands on green Huelo Point overlooking the ocean. The large basaltic boulders that form its walls are covered with buff plaster. The combination of setting with the building's thick walls, French windows, and towering steeple suggests the rugged backdrop of Wuthering Heights; the architecture seems more suited to a temperate climate than to the warmth of the tropics. Tiger lilies, asters, roses, and ti plants grow in the churchyard, and you cross the fence on a stone stile.

Hoalua Bay is on the far side of Huelo Point. Projecting into the bay is a pyramid of rock 50 meters high, Pu'u Ka Ae.

Kaulanapueo Church.

Left: Hoalua Bay. Projecting into the bay is Puʻu Ka Ae, still attached to the land, but well on its way to becoming a sea stack.
Right: Keōpuka Rock near Honomanū Bay is a sea stack now isolated from the mass from which it was carved. Haipuaʻena Stream falls 80 meters into the sea at Mōiki Point.

Presently it is part of the shore between Hoalua and Hanawana streams. In recent geologic times, when the ocean was at a higher level, it may have been a sea stack, just as Moku o Kau, a half kilometer to the east, is today.

For the next 6 kilometers, the seacoast is remarkably

Two-pronged Makaīwa Bay interrupts the straight coastline of sea cliffs between Waipiʻo and Honomanū. The lighter areas in center are cultivated fields; Hāna Highway runs along the upper part of the scene.

straight, with only slight indentations. Sea cliffs face directly into the northeast trade winds, and streams fall 120 meters into the sea. Halfway between Waipiʻo and Honomanū is Makaīwa Bay. Two projections of Honomanū rock, one on either side of Oʻopuola Stream, are conspicuous prongs in an otherwise straight seacoast. Makaīwa means 'mother-of-pearl eyes', as in the inlaid eyes of a *kiʻi* image.

The road winds through groves of mango, *kukui*, eucalyptus, and bamboo, with greens varying from nearly white to those of darkest hue. At Waikamoi, the stream bank above the road is steep and its boulders are well rounded. Pools are close to the old cement bridge, and water drips out of rock layers, adding to the stream's volume. The Waikamoi Ridge Trail passes through a stand of eucalyptus to a grassy area, then to an overlook of forested valleys 1000 meters distant. A rise of 100 meters in the trail gives the hiker a view down upon the Kōlea Reservoir.

Keōpuka Rock, a double-humped, 40-meter sea stack, stands outside Puohokamoa Bay. The waterfall behind Keōpuka Rock at Mōiki Point drains Haipuaʻena Stream. The road in this region makes a sharp hairpin turn about a kilo-

meter inland, and passes in quick succession the falls of Puohokamoa and Haipua'ena, then runs out to the sea to begin
its journey along the high cliffs.

Between Haipua'ena Falls and Honomanū Bay is a flat
region of Kaumahina State Wayside Park. Here is a fine vantage point from which to enjoy the Ke'anae scene. A nature
trail offers hours of pleasant sauntering among *hala* trees 10
meters tall, standing on their stilts. Signs acquaint you with
both the native and the exotic vegetation. Read their Latin
names, syllable by syllable, and you'll find yourself intoning
a chant consonant with the solemnity of the place:

Nephrolepus exaltata
Jacaranda acutifolia
Eucalyptus robusta
Aleurites moluccana
Hibiscus arnottianus
Melaleuca leucadendra
Pandanus ordoratissimus

Honomanū Bay is a 500-meter indentation in the coastline. The road descends a 60-meter cliff to cross the flat allu-
viated valley bottom. The walls of Honomanū Valley increase in height to 350 meters upstream, swelling into an
amphitheater 4 kilometers from the ocean. In the walls of
this canyon are the oldest exposed rocks of Haleakalā—thin-
bedded basaltic lavas of the Honomanū Volcanic Series. The
great volcanic edifice that is Haleakalā was built of Honomanū volcanics. Later, these older flows were covered with Kula
and Hāna volcanics, a covering so complete that the oldest
rocks in this ancient shield are found only here and in a few
other places.

Several types of streams flow down the slopes of Haleakalā
—young ones that barely indent the surface, intermittent
ones that run only during heavy showers, persistent peren-
nials that cut deep V-shaped valleys, and mighty ones that
have gouged out huge canyons like Honomanū.

Erosion of a land mass begins with a swift-flowing stream

cutting a gully. Time goes by. The deepening and enlarging of the gully continues at an increasing rate as springs and seeps in the valley walls are tapped, adding to the stream's volume. Eventually, the enlargement of the valley produces a broad canyon with an amphitheater head, such as Honomanū, where rainfall is plentiful.

The road between Honomanū and Nuaʻailua hangs on the edge of a cliff 60 meters above the sea. *Hala* trees cling to the cliff and overhang the ocean, and vines reach out with curling tendrils searching for support. Ahead, jutting out into the sea, is the low, broad peninsula of Keʻanae.

KEʻANAE

Keʻanae, 'the mullet', is a wide valley with a flat floor. Once it was narrow and deep, carved during a long period of erosion that followed the completion of the Haleakalā shield.

In ancient times, lava flowing down Koʻolau Gap from summit eruptions created Keʻanae Peninsula.

You can estimate the depth by projecting the valley walls down to an approximate stream bed. Tapping large amounts of ground water as it worked its way headward, Keʻanae formed an amphitheater in the heart of the mountain. Then a new series of eruptions at the summit poured lavas of the Hāna Volcanic Series through Koʻolau Gap and down into the valley, flattening its floor and spreading out into the sea as a fan—the Keʻanae Peninsula.

The lava flowed over conglomerate that had formed when the sea was 30 meters higher. It plastered valley walls and poured into very deep valleys that had been cut during a time when the sea was 20 meters lower than it is today. Washed downstream to Keʻanae from a source up in the mountain is a fine-grained volcanic rock. Red in color and polished with a "lacquer" of iron-manganese hydroxide, the stones are known as *pōhaku koko*, 'bloodstone'. A sample is in the Bernice P. Bishop Museum collection.

Taro patches covering the Keʻanae Peninsula glint in the sun like mirrors, and palm trees grow along the beach in this beautiful garden region. Rectangular in shape, the taro patches are separated from one another by green lanes. Ponds in which plants are just appearing are light green; those with mature plants are dark; and the unplanted paddies reflect the blue of the sky.

Taro, *Colocasia esculanta*, was brought from Tahiti by early Polynesian voyagers. A staple food in their diet, it contains calcium oxalate crystals which irritate the throat if the taro is eaten raw. Cooking and pounding, however, destroy the crystals and make good, nutritious poi.

The Keʻanae Arboretum trail travels 2 kilometers up Piʻinaʻau Stream past a variety of types of taro. Sections of the arboretum are devoted to native forest trees, introduced tropical trees, and cultivated Hawaiian plants.

It is 2 kilometers from Keʻanae to the village of Wailua. Between them is Pauwalu Point. Statuesque Moku Mana islet, many sea stacks, caves, and a natural arch make it a lively place, and the small inlet beside the point has the

Taro fields at Ke'anae.

bright-sounding name of Hahaha Bay. Waiokamilo Stream makes a right-angle bend at Pu'u Ililua and drops 30 meters to the sea at Waiokilo Falls.

The road meets Waikani Falls just beyond Ke'anae Valley Lookout, and the stream empties into Wailua Bay. A triangular sea stack just beyond Wailuaiki Bay marks Papiha Point. From here to Nāhiku is a stretch of sea cliffs like those near Honomanū Bay. The road, though, is not hanging on the edge; it lies far inland, a thousand meters from the coast.

Streams are closely spaced on windward Haleakalā. Hanawī, Makapipi, and Kūhiwa streams all enter the ocean within 2 kilometers of each other at Nāhiku. A trail from the road at Hanawī Gulch leads down to the Big Spring, the largest spring on Maui. Each day, 40,000 cubic meters of water pour out of it—enough to flood a football field to a depth of 8 meters.

Sea stacks as seen from a cave near Pauwalu Point.

Sea cliffs disappear at Nāhiku. Deep valleys, wide canyons give way to small gullies and shallow valleys. Gone is the rugged, deeply dissected topography. But there's plenty of rainfall here to carve canyons. Why the difference?

The change occurs at a place where we encounter rocks of a vastly different age. The surface over which we have been traveling is made of rocks of the Kula Volcanic Series; the land is old and deeply dissected. But here at Nāhiku we meet rocks of the younger Hāna series that cover this eastern end of Maui all the way to Kīpahulu.

Hāna lavas flowed over the old Kula volcanics, filled in valleys, smoothed the surface, and built a new shoreline. Highly fluid, the Hāna flows filled some valleys to a depth of 15 meters or more. At other places, the still-hot lava drained

Sea stacks at Pauwalu Point, with Hahaha Bay in the foreground. A long stretch of sea cliffs ends at Makapipi Stream, and the Nāhiku region begins. Cones on the skyline mark Haleakalā's east rift.

quickly toward the sea, leaving a half meter of rock stuck to the valley walls. Such "plastering lavas" are prominent in Keʻanae Valley as well as at Nāhiku.

A few rubber trees are found in Nāhiku. Around 1898, the world supply of rubber came from wild trees in Brazil, Africa, Malay, and a few other tropical countries. An ever-increasing demand was pushing up the market price of rubber. Entrepeneurs in Hawaiʻi, hoping to enter the market, planted more than 26,000 rubber trees of three varieties in the Nāhiku region, where they grew rapidly and well. After five or six years, the larger ones were tapped and found to produce a latex of excellent quality. The project looked to be so promising that the Nāhiku Rubber Company was formally organized in 1905.

New plantations were also maturing in other parts of the world where labor and production costs were lower than in Hawaiʻi. Faced with higher costs and a subsequent decline in the market price, the Hawaiian venture proved to be uneco-

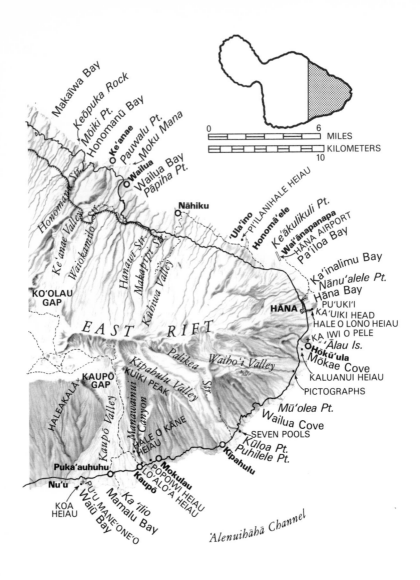

nomical, and so the first rubber plantation on American soil
failed.

A depression on the smooth side of the mountain above
Nāhiku, at an elevation of 1200 meters and about 8 kilometers from the sea, is Kūhiwa Valley. It is not part of any defi-

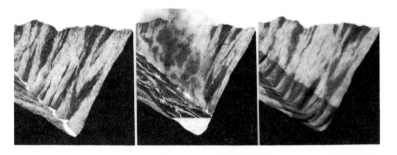

Diagram showing the formation of "plastering lava." *Left:* A stream cuts a deep valley. *Center:* Flow of fresh lava from a higher eruption floods the valley. *Right:* As the flow drains out of the valley, residual plastering lava clings to the valley walls.

nite valley down below; rather, it is the head of a former amphitheater that has now been almost completely buried by Hāna flows. Because Kūhiwa is a wet region, it is no wonder that an amphitheater formed there. The rain gauge station records an average of 9.8 meters of rain a year, making it the second wettest place in Hawai'i, surpassed only by Kaua'i's Wai'ale'ale.

The coastal plain widens at 'Ula'ino, 'stormy red' and we come to a historic region of ancient Hawai'i. The area is isolated, separated from Wailuku by 50 kilometers of sea cliffs. Because of the geographical barrier, this region of Maui has in the past been more closely associated politically with the island of Hawai'i than with the other parts of Maui.

Honomā'ele is the site of the largest *heiau* on the island, Pi'ilanihale. The stone platform, 105 meters by 128 meters, is terraced in several steps on two sides. On its north side, five step terraces raise the top of the platform to 17 meters above its base. No other *heiau* in the islands is so extensively terraced. Small pebbles pave its surface, and a lava block trail nearby may be part of the Kiha a Pi'ilani Trail that is supposed once to have circled the island. Here at Pi'ilanihale, 'home of Pi'ilani', was probably the royal abode of the great Pi'ilani family of Maui chiefs in the sixteenth century.

An amphitheater marks the head of the former Kūhiwa Valley. Later lava flows *(shown in dark)* filled the lower reaches of the valley, isolating the amphitheater.

The sea washes a stand of **basaltic columns** at Keʻākulikuli Point near the Hāna airport. Hexagonal in shape, the columns are about a half-meter thick and 8 meters long, pink at the water's edge with algae and *'opihi*. Such basaltic columns form where lava cools slowly, particularly where it has ponded to great depth in valleys. Shrinking as it cools, the lava develops tension cracks. The distribution of the cooling centers is such that six-sided columns tend to form, although five- and seven-sided columns occur about as frequently. Waves climb Keʻākulikuli Point and roll along the jagged tops of the columns, while tons of ocean water surge through caves four times a minute.

The sand is black at Paʻiloa Bay, a sharp indentation in the coast beyond Hāna airport. Water-worn pebbles and cobbles clatter as they roll back and forth with the movement of the waves. You can also hear the booming of the sea as it compresses—then releases—air in lava tubes and caves.

The caves of Wai'ānapanapa, 'glistening water', are reached by a trail through *naupaka* and *hala* forest. The dry upper chamber of this double cave was formed when loose clinker fell from the wall; the lower cave is a lava tube filled with water—a tube through which you have to swim in order to reach the dry chamber.

Legend tells us that a Hawaiian princess had a jealous husband who became angry with her when he thought she was bestowing her favors upon another. To escape his wrath, she and her maid hid in Wai'ānapanapa Cave. They entered the water-filled lava tube and swam under a rock ledge to the higher dry cave where they rested and waited, thinking themselves safe. After searching far and wide, the princess's husband entered the cave. She and the maid remained quiet, but reflections of her *kahili*, the bright royal emblem, shone on the mirror surface of the pool, revealing their hiding place. The husband killed both women, and the red water that

Platforms of house sites near Honomā'ele are built of aa boulders. Offerings of ti leaves are carefully wrapped around beach cobbles.

Basaltic columns in a dense aa flow near Keʻākulikuli Point. *ʻOpihi*
cling to the columns and waves surge in and out of the sea cave.

sometimes appears in the cave represents the blood of the vic-
tims of the man's rage.

HĀNA

Hāna: *ʻāina ua, lani haʻahaʻa,* 'rainy land, low-lying sky'.
The coastal plain narrows and we have reached the eastern-
most point on the island. A line of cinder cones running up
the mountain and into the clouds marks the east rift zone of
Haleakalā.

Waves roll into Kaʻinalimu Bay and tumble up onto the
cobble beach. Young surfers riding the breakers hoot and

Hāna. The town lies behind Kaʻuiki and Puʻukiʻi. Hāna Bay is at right, and cinder cones on the flank of Haleakalā indicate the east rift.

scream like birds. A quiet fisherman, net draped over his shoulder, moves slowly, looking intently into the water. When he finds the right place, he hurls the net; its weights peel out and spin the net into a beautiful circle before it settles on the sea.

What was it like here a thousand years ago, when his distant ancestors arrived from Tahiti? New forms of vegetation came with those ancient immigrants—taro, ti, coconut, breadfruit—and they brought with them old ways of making music as well: singing, dancing, the drum and nose flute. Since their arrival, how many times has this region been the target in the attempts of powerful chiefs from the island of Hawaiʻi to unite the islands under a single command? The problem must have been faced many times: allegiance to whom?

Nānuʻalele Point, curving far out into the sea, is the division between the bays of Kaʻinalimu and Hāna. A rough trail along the beach crosses old aa flows. Tidal pools along the sweep of the point move with the sea, and ancient flows arch across them as natural bridges. Beyond the point is crescent-shaped Hāna Bay, a 900-meter bite at the extreme eastern end of Haleakalā.

Hāna Bay was once called Kapueokahi, 'the single owl'. Guarding the southern end of the bay is Kaʻuiki Head, and beside it is the small islet Puʻukiʻi, 'image hill'. A lighthouse on its 22-meter summit is a vantage point from which to view the Hāna region. The old name for this islet was Puʻukū, 'upright hill'.

Kaʻuiki Head is a large hill at the edge of the sea, a late cinder cone on the east rift zone of Haleakalā, formed during the Hāna Volcanic Series. Tree-covered and rounded, it is a disjuncture in a land of sloping meadows. A sizeable portion of the cone has been worn away by the sea, exposing its internal structure. The red color is due to oxidation of its iron-rich lavas. Black ash, which fountained from Kaʻuiki as well as from other cones along this east rift, is scattered all over the Hāna region.

Because of its thick cover of vegetation, 188-meter Kaʻuiki Head is difficult to climb. Bunchgrass covered it in ancient times. But with the introduction of exotic plants the endemics lost out, and the hardy modern vegetation is difficult to get through. Nor is the panorama from the top that good; ironwood trees block the view.

Many battles have taken place here at Kaʻuiki, for it was a fortress for many years. Kalaniōpuʻu, king of Hawaiʻi, captured it in 1754; he held it until 1775, when he was defeated by Kahekili, king of Maui, at Kaupō. Kahekili reconquered East Maui by cutting off the water supply to Kaʻuiki Head. He baked the bodies of its defenders in earthen ovens to show his contempt.

A cave at the base of Kaʻuiki Head near the islet of Puʻukiʻi is the birthplace of Queen Kaʻahumanu, favorite wife of Kamehameha I. On the opposite side of the cone is Hale o Lono *heiau.*

A broad expanse of sea and sky is revealed at Hāna-of-the-low-lying-clouds. Moisture-laden trade winds roll up long lines of clouds which, mixed with patches of blue, break over Hāna-like waves.

Here at Hāna all is simplicity and quiet. Meadows surround the village. The road is rough, bordered by wooden fences interrupted by occasional lanes. The scene is pastoral. The pace of living is slow and dignified—old churches, a Buddhist temple, a girl walking to church with a Bible in her hand. The Hasegawa General Store and the Ranch Store take care of the material needs of the people. Hāna, a rustic place isolated from the rest of Maui . . . from the rest of the world.

Ka Iwi o Pele, 'the bone of Pele', is a 140-meter cinder cone partially worn away by wave action. Its sides are vertical, exposing the layered beds of ash. *Naupaka* holds the leeward side of the cone, ironwood holds the windward. A thousand meters offshore is 'Ālau Island, 'many rocks', believed to have been formed by Pele. Ocean waves crash against the island, sending spray all the way to the top of its 50-meter summit, which is crowned by two palm trees.

South of Ka Iwi o Pele is the largest of five fish ponds and traps built by the old Hawaiians. It is a low wall of stone about 180 meters long. Waves break over the wall at high tide, and fish are swept into the pond. But as the tide lowers, the fish are trapped within and easily caught. An opening in the wall also allows fish to enter, and a net placed across it prevents them from escaping.

Ka Iwi o Pele was a busy place in ancient times. Here the pig god Kamapua'a ravished Pele. Defeated in a battle with her older sister, Nāmaka o Kaha'i, Pele left her bone here. From the top of this cone, the god Lono-muku made his ascent into the sky to go and live on the moon. It also serves as a landmark for a fishing station off Kīpahulu. Most important, it was near here that the demigod Māui raised the islands out of the sea.

Māui used a great hook to fish up the islands. To give the hook supernatural powers, he baited it with the *'alae*, a mud-

Center: Ka Iwi o Pele; in the foreground is ʻĀlau Island.

hen. The *ʻalae* was a great benefactor because it brought fire to the people of old. But it did so at great risk to itself and was transformed in the very act: the bird's forehead, once white, was scorched in the intimate encounter with the fire and is now a brilliant red.

Baited with the sacred *ʻalae*, Māui's hook had divine

power, and he was ready for another of his heroic deeds. Pushing out from shore in a canoe, he and his three brothers sailed into the open ocean until they came to a place he knew to be right. There he swung the magic hook baited with the sacred bird round and round over his head, letting the rope gradually play out in an ever-increasing circle of power. Then he let the hook fly. High into the air it sailed, far out over the waves. Striking the surface with a mighty splash, it quickly disappeared from sight.

Down and down it sank, down into the depths of the sea. "Paddle!" shouted Māui. "Paddle as hard as you can. But don't look back!" They paddled—paddled as they had never paddled before—for something very unusual was on the end of the line.

Caught by the magic hook, a large piece of land began to emerge from the water. But curiosity rose with it, and the impelling urge to see what had never been seen before was overwhelming. One of the brothers looked back.

The line snapped. The magic was gone. The illusion was shattered and the land mass with it, breaking into many pieces.

You can see those pieces in the sea today as the Hawaiian Islands. And you can see the great magic fishhook in the sky —Manai'i-kalani, 'made fast to the heavens'—a great group of stars lying along the path of the sun for all the people on Earth to admire.

But not all the Earth's people see that star group as Māui's fishhook. For most, it is Scorpius the scorpion, clearly visible as one of the finest of the zodiacal constellations. Brightest in Scorpius is Antares, the "rival of Mars." The Hawaiians know it as Hōkū'ula, 'the red star'. It may well be that this giant star in Māui's fishhook gave rise to the name of the village at the base of Ka Iwi o Pele—Hōkū'ula.

The soil at Hōkū'ula is also red—almost maroon—and mixed with black lava. A platform of rocks in an open field is Kaluanui *heiau*. The lower platform, smaller than the upper one, was used for tapa drying, and the *heiau* is supposed to have been a shrine to the goddess of tapa beaters. Trees growing through the ancient ceremonial site dislodge the rocks and the *heiau* deteriorates.

A *hala* forest grows on a bluff above Mōkae Cove. It is a pleasant experience to walk here, for the floor of the grove is soft, covered with a natural matting of *hala* leaves, softer, even, than a finely woven *lauhala* mat.

Waihoʻi Valley, 'returning water', begins at Mōkae Cove. The first of three big valleys on the southern side of Haleakalā, its stream is 8 kilometers long. The stream cut a valley 600-meters deep during a great erosional period. Later, Hāna lavas flooded in, covering the valley floor to considerable depths.

Flowing down the center of Waihoʻi Valley is Waiohonu Stream ('water of the turtle'). Along its banks are pictures painted hundreds of years ago by unknown artists. A few hundred meters of boulder hopping in the stream will bring you to a constriction in the valley where the walls are vertical. By swimming or wading past this narrow section you can get farther upstream to the pictographs. It is startling to come suddenly upon them—to come face to face with the

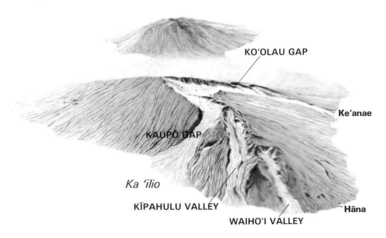

Maui from the southeast. The vertical scale is greatly exaggerated.

Typical pictograph figure.

work of an ancient rock painter, to see the symbols of his reality displayed so prominently in this quiet glen.

Unlike petroglyphs, which are pictures pecked into the rock with stone tools, these **pictographs** are drawn with red dye on slabs of rock wall. Simple and basic, the figures have long necks and long torsos. Spotlighted in the sun and illuminated by light reflected off the rock walls, the stately images in this silent grove lend a feeling of the presence of ancient man as well as insight into his mind.

WAILUA AREA

Papahawahawa Stream flows along the western side of Waihoʻi fan and empties into the sea at Mūʻolea. From the ancient ruins of *heiau* and house sites at Mūʻolea to the basaltic columns of the Seven Pools is an indented region of 60- to 100-meter sea cliffs and spectacular waterfalls.

Wailua Cove is the prominent shore feature in this pie-shaped region between Waihoʻi and Kīpahulu Valleys. The sector is composed of old lavas of the Kula Volcanic Series. An erosional remnant of the Kula shield with a thick covering of soil, it was too high to be flooded with later Hāna lavas. The heads of Waihoʻi and Kīpahulu are close together near the top of Haleakalā. Due to the radial drainage pattern, the streams flow in different directions—Waihoʻi almost due east, Kīpahulu southeast; and a 4-kilometer line of sea cliffs separates them at the base of the sector.

The region is moist and cool. Water drips out of rock layers, and many waterfalls are found here. The 30-meter Wailua waterfall is near the cement bridge covered with chartreuse-colored moss. The most spectacular waterfall is the 120-meter Waihiʻumalu Falls.

Pīpīwai Stream, ʻsprinkling waterʼ, flows along the western side of the Wailua area. Above the Seven Pools it joins the adjacent Palikea Stream in Kīpahulu Valley, and together they flow in ʻOheʻo Gulch to the sea at Kūloa Point. A hike on the trail above the Seven Pools through guava, overgrown patches of taro, and past a bamboo grove leads to Waimoku Falls in Pīpīwai Stream.

KĪPAHULU VALLEY

The second of the three large valleys on the southeast side of Haleakalā is wide, flat Kīpahulu, ʻfetch from exhausted gardensʼ. Its sides are steep and at its seaward end is a 4-kilometer fan.

Hexagonal basaltic columns in the bottom of ʻOheʻo Gulch on the Kīpahulu fan look like cobblestones fitted closely together. Water flows swiftly in the stream, bouncing off rock walls, turning inside out, churning and eddying, wearing away at the fracture and jointing planes. A trail leads from the road past 30-meter walls which contain cascading torrents, then through a quiet grove of pandanus. On the grassy meadow by the sea are stone platforms as well as the walls of ancient house sites, indicating that this beautiful region was a favorite of people long ago.

Fortunately, the Seven Pools and a large part of Kīpahulu Valley are preserved for posterity as part of the National Park System. The preservation of the land was made possible by the Nature Conservancy Group which raised the money to purchase the valley in 1969.

The upper part of Kīpahulu Valley is of great interest to botanists. Remote and difficult to reach, it abounds in rare flowers, birds, and endemic vegetation—forms found nowhere else in the world. Hawaiian bunchgrass, typical of the soil cover in many parts of the island before the advent of plantations and cattle ranching, grows at an elevation of 2500 meters. Grasslands hold water like a sponge, insuring good water supply for towns and villages.

Kīpahulu Valley has had a complicated geologic history. Long after the Honomanū shield was built, giving Haleakalā its present form, Kula activity started. Abundant flows of lava at this time covered the Honomanū shield entirely. A long erosional period followed. Heavy rains running off in streams over many thousands of years of time carved deep valleys, and one of the streams excavated a huge amphitheater.

Activity of the Hāna Volcanic Series began. Lavas poured

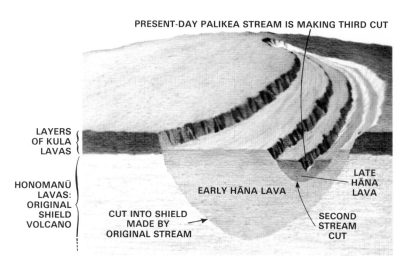

Cut-away diagram showing stages in the formation of Kīpahulu Valley.

down Kīpahulu Valley from a point near the summit, filling it to a depth of more than 600 meters, flattening the floor. Streams began cutting into these early Hāna flows, cutting a valley 1000 meters deep and 1500 meters wide. The new V-shaped channel was cut along the eastern edge of the flow, leaving a series of terraces.

Renewed activity once more poured lava down Kīpahulu Valley, partially filling the incised canyon. A series of three levels now forms the valley floor. Palikea Stream is currently cutting into the most recent flow.

The climate of Kīpahulu is similar to that of Waihoʻi and Hāna. Prevailing winds blow almost parallel to the coast. Annual rainfall may be as much as 200 centimeters a year at lower levels, increasing in the heights.

Prized by the *aliʻi* for their productiveness, these lands supported a large population. Because of their value, warfare was common, particularly with Hawaiʻi, only 50 kilometers distant across ʻAlenuihāhā Channel. Kamehameha I stopped here on his way to conquer Maui and built a *heiau* at Maʻulili.

After the abolishment of the *kapu* system in 1819, the region drifted into obscurity, commerce then centering around Lahaina. The sites and structures near Puhilele Point that still remain are those of a Chinese peanut-growing community. Big ranches have acquired most of the land. Stone walls are cattle fences—boundaries of recent origin. Here at Kīpahulu, where thousands once lived, now only a few families remain.

MOKULAU AREA

The broad undulating land of the Kīpahulu lava fan ends at Kukuiʻula, 'red light'. A 5-kilometer stretch between here and Kaupō Valley is another triangular region of Kula lava, just as at Wailua—sea cliffed at one end and sharp ridged at the other. Also just as at Wailua Cove, this area was too high to be flooded by younger Hāna lavas, and so it represents the surface of the old Kula shield. The top of the sector is the peak Kuiki, 2302 meters, near Palikū Cabin in Haleakalā Crater.

High on a hill overlooking the sea at Mokulau, 'many islets', is Popoiwi, one of three great *heiau* attributed to Chief Kekaulike. Measuring 50 by 100 meters, Popoiwi was a *puʻuhonua*, a place of refuge, built in about 1730 when Kekaulike was at the height of his power. Later he was defeated by an invading army from Hawaiʻi in a raid that began a half century of warfare between the islands of Hawaiʻi and Maui.

Scooped out of the smooth slopes of Haleakalā is the deep canyon called Manawainui, 'large water branch', a large

Manawainui Valley.

amphitheater-headed valley that seems to go nowhere. Streams, pouring over the sharp rim at an altitude of 1400 meters, fall long distances into plunge pools, overflow these basins, and cascade to still lower pools. When seen from a distance, this succession of falls looks like a disjunctured ribbon of water.

Despite its large catchment area, Manawainui Stream is sometimes dry, for water sinks quickly into the ground in this arid region. Magnificient in its upper reaches, the valley seems to taper almost to nothing at its mouth. Lava flows from adjacent Kaupō Valley pushed over the end of Manawainui Valley, smoothing it over and diverting the stream to the east.

Valley walls are steep; the canyon is narrow and deep. Rock is loose, and sometimes animals dislodge boulders, sending them bouncing off walls and crashing into the vegetation at the bottom; so getting to the end of the valley can be hazardous.

The crested honeycreeper and the "happy-face spider" live in this region of Haleakalā. So do wild goats and huge pigs. The small black pig which the ancient Polynesians brought with them constituted no great threat to the grasslands and forests. But with the importation of pigs from other parts of the world and the inbreeding of that animal with the Polynesian pig, the result is a large animal with huge tusks that roots through the soil like a bulldozer, literally upsetting the vegetation in its search for food. Young seedling trees have little chance of surviving the onslaught of pig and goat; and so the forest recedes.

Manawainui Stream flows past the three *heiau* built by Kekaulike—Hale o Kāne, Lo'alo'a, and Popoiwi. Closest to Kaupō is Lo'alo'a, 'the pitted'. Long and narrow (155 by 30 meters) it is a rough open platform of lava rocks. A wall extends across the split-level surface of the platform, probably separating the ceremonial area from the secular. Three step terraces support the southeast side, and four the northeast side, to a height of 10 meters.

KAUPŌ

Kaupō, 'the night landing place of canoes', is the biggest of the three large valleys on the southeast side of Haleakalā. Once a deep valley, it has since been flooded with late lavas that flattened the floor and pushed out into the sea. The lava fanned out to create a peninsula whose shoreline is about 8 kilometers long—from the village of Kaupō westward to Nu'u.

"Kaupō is indeed a green land and so is Hāna. They look so open and pleasant to live in because the wind is always blowing," was Thomas Manapau's description of the region when he visited there in 1921 with Dr. Kenneth P. Emory of the Bishop Museum. A resident of Kaupō, Mr. Ahuli'i, told them of the local winds and rain:

Moae is the customary wind that blows strongly but pleasantly from the sea and sometimes from the land. It is sung about:

> *Where are you, O Moae wind*
> *You're taking my love with you.*

Moae-ku is like the moae but much stronger. This wind is said to have been born in Hana, grew up in Kipahulu, attained maturity in Kaupo, become aged in Kahikinui, grew feeble at Kanaio, rested and let its burden down at Honoaula. Here is a song for this wind:

> *Where are you, O Moae-ku*
> *You make much work on a stormy day.*

Kulene wind comes with rain. It is strong and blows out to sea from the land.

Kaomi is a strong, blustering wind whose strength did not last long but blew a gentle pressure. It is sung thus:

> *The wind blows in a gale,*
> *Then it gently presses*

Kiu is a wind that flies along and seems to sneak by to the mountain of Haleakala. Here is the song of this wind:

> *The Kiu is the wind*
> *That lives in the mountains.*

Naulu. This wind goes with the naulu clouds:

> *The Naulu is the wind*
> *It bears the naulu clouds along.*

Makani kaili aloha o Kipahulu. The love-snatching-wind-of-Kipahulu is the usual Kipahulu wind. It blows down from the mountain and goes out to sea.

And of rains, Manapau writes: "The names of the rains of this very famous place are pretty, and I do not think that there are rains anywhere else to compare with these." He describes them thus:

Noenoe uakea o Hana is a misty rain and white. It comes in the morning and ends as the morning waxes. This rain had often been sung about by singers, more so when a glassful had been taken and the mist thickened so that one's companions could hardly be distinguished. That was the time when a man felt very smart and his singing more joyful as he sang the song of the rain:

> *Misty and white is the rain of Hana*
> *Companion of the Malualua wind.*

Hana ua lani haa haa (Hana of the low rains from heaven) is described thus: A low-hanging cloud comes from the ocean, then the rain falls; that is why it is so named.

Ua Awa is a dark cloudy rain that falls all day in the mountains.

Ua Koko is a rain that spreads over the surface of the sea like a rainbow. It is a sign of trouble of some importance.

Ua Naulu moves over the mountain on a clear day with a naulu cloud.

Ua Noe is a light shower and mist that remains in kula lands.

Ua Haleu Ole is a naughty rain. When one wants to relieve nature, that is the time it comes with such suddenness and clears up just as suddenly and then falls again. That is why it was called "wipeless." The writer thinks that it is a modern name, and because of a lack of a better one, it was given the mischievous name.

Ua Lilinoe o Haleakala—rain of Haleakala—a famous rain belonging to that mountain.

Stone shelters are common along old garden plots. The walls, *ka ua pe'e pa pōhaku*, are about 2 meters high and 3 meters long; they provided shelter from sudden rain squalls that drive across the land at certain times of the year.

A trail from Kaupō Village, along the west side of Manawainui Gulch and past the Hale o Kāne *heiau*, leads to the top of Haleakalā. The hike is extremely difficult because the trail is rough and steep, and the climb is on the south side of the mountain, which is exposed to the full force of the sun. Water is scarce. The 13-kilometer trail climbs to an altitude of 1950 meters—to Palikū Cabin in Haleakalā Crater— where it connects with the trail across the crater floor. Hikers more often walk down from the top after staying in Palikū Cabin, but that, too, is a rugged venture.

Of the largest of Kaupō's points, Ka'īlio, 'the dog', Thomas Manapau writes, "Perhaps it was its resemblance to a dog that gave it its name, or perhaps it was named for a supernatural dog of that place." Chief Kahekili of Maui successfully stopped the invading army of Kalaniōpu'u at Ka'īlio; young Kamehameha was in that defeated army.

Capping the 25-meter peak of Ka'īlio is a limestone layer a half meter thick. On it are shells of a marine gastropod, *Littorina*. The cap is not part of an ancient emerged reef; it is actually forming today as sea snails climb the point and their shells are incorporated into the limestone.

The road passes through the junction at the middle of the fan, Puka'auhuhu, ' *'auhuhu plant hole*', paralleling Mamalu Bay about 1000 meters inland. Near the western side of the fan is Pu'u Mane'one'o, *itchy hill*, a 108-meter hill seacliffed at Waiū, 'female breast'.

Lava flows surround Pu'u Mane'one'o, but the prominence itself is not built of lava; it was formed of mud and boulders that rushed down Kaupō Gap. Heavy, persistent rains high on the mountain saturated a vast accumulation of volcanic ash and debris, which thus became very unstable. Earthquakes rocked the island, shaking the material like a bowl full of jelly. Set in motion and moving under the force of gravity, the slurry of fine material mixed with water rushed down the slope at a speed of 30 to 50 kilometers per

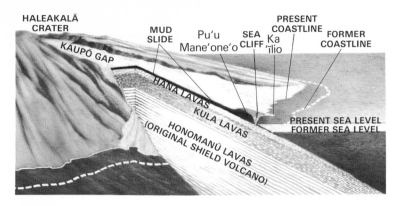

Cut-away diagram showing structure of Kaupō Valley. Not to scale.

hour; lubricated by interstitial mud, the melange cascaded down old stream courses and swirled around the bends, picking up anything moveable—trees, volcanic bombs, boulders several meters in diameter. The entire mudflow is more than a hundred meters thick, and at the time of the flow the sea was 100 meters lower than it is today.

The road swings around Puʻu Maneʻoneʻo, very close to its base, and you can see the jumbled, chaotic mass of the mudflow. In it are angular and subangular blocks of basalt and andesite, some stained red. No sorting. No stratification. What a contrast this mudflow is to the stream-sorted conglomerates in the nearby valley walls!

Atop Puʻu Maneʻoneʻo is a *heiau*, a large open platform of rough lava with several enclosures.

Kaupō Valley is floored with Hāna lavas, but the valley walls are those of the Kula Volcanic Series. The V-shaped canyons beyond Kaupō are cut in Kula lavas; their mouths are covered with Hāna flows and their streams diverted by that more recent activity.

The view of Haleakalā from Puʻu Maneʻoneʻo, in the words of Manapau, is "regal and majestic. Mists surrounded the summit. Our companion pointed to a spot on Haleakala

Close-up of Puʻu Maneʻoneʻo showing the jumble of material that was carried down Kaupō Gap by the ancient mudflow.

called Pauli nui a Kane, *Kane's big depression*; and to Maupaakea, *thick white snow*, a spot often mist-covered and snow-covered sometimes; and to Kolekole, *raw*, a barren raw-looking place high up on the mountain about ten thousand feet from sea level. Nuu," continues Manapau, "was a cattle shipping port and landing place for fishermen in the olden days."

Near the village of Nuʻu is a petroglyph site containing pictures of people and dogs. Salt pans at Nuʻu were made by chipping shallow depressions in boulders. ʻAlalākeiki, 'child's wail', is the name of the cave near the road; it is also the name of the channel between here and the island of Kahoʻolawe.

The economy of Kaupō and Kīpahulu must have depended

heavily upon agriculture. Shores are rocky and cliffed; reefs do not exist, so inshore fishing is limited. Though the Hawaiians were skilled in handling canoes, there are few places in this region where landings could be made safely. Canoe sheds are rare. The few that are found are located chiefly at stream mouths and at a few sheltered beaches at Mokulau.

Kīpahulu was ideal for the cultivation of taro because of the abundance of water. At drier Kaupō, the crop was sweet potatoes, which do not require extensive terracing or irrigation systems. Rainfall diminishes from here toward the west. Though the land looks severe, it was heavily populated at the time of Captain Cook.

BEYOND KAUPŌ—THE "DISMAL COAST"

Such a coast it seemed to La Pérouse in May 1786; he wrote of it in his *Voyage around the World:*

> . . . to add to our mortification, we did not find an anchoring place well sheltered till we came to a dismal coast, where torrents of lava had formerly flowed, like cascades which pour forth their waters in the other part of the island.

The country is dry and rugged. Austere. Abundantly populated in ancient times, it is now abandoned. *Heiau* and house sites are reminders of activity in that time so long gone. Even more recent structures are almost completely deserted. Small churches, built a century ago, hold only occasional worship services.

The land is desolate, but amidst these relics the past is strangely alive. A thousand or more people must have lived in the region between Nu'u and Kiakeana Point. More than 140 house sites still remain in that 13-kilometer stretch of seacoast—an average of nearly one every 100 meters.

House sites and villages are generally found at the mouths of gullies, well within sight of the sea. Built on rocky, rough land near the shore, a village is a collection of large and small house sites, canoe sheds, storage caves, animal pens, burial

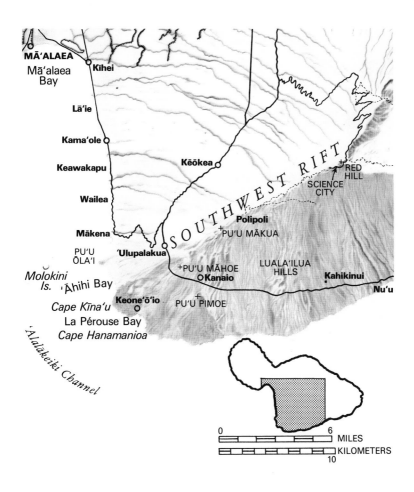

sites, and *heiau* platforms. The better land, smooth and covered with soil, was preserved for agricultural use.

Villages near the sea had the persistent problem of brackish water. But villages high up on the slopes had their water problems, too. Small streams and rivulets, some of them running only intermittently, were dammed for a supply of fresh water. The large village of Kahikinui, 'great Tahiti', used such a system. Built on a prominence between two gulches at an elevation of 475 meters, it lies 3 kilometers from the sea.

Kahikinui as it may have looked when a thriving settlement. In the foreground two men are playing *kōnane*; houses and work areas are shown beyond.

Kahikinui is an amazingly large settlement in this desert country. It climbs the hill in a series of step terraces for tens of meters. The eastern side of the terracing is broken, but its western portion is still intact. Small pebbles covered the floor of the village, and within enclosed areas are *imu* pits for cooking. Now the platform is rocky and barren, but once Kahikinui had its windbreaks of stone walls, houses of thatched roofs, and stony floors made comfortable with thick *lauhala* matting.

From the high vantage point of Kahikinui, you can look to the south across 'Alenuihāhā Channel to the snow-covered shields of Mauna Kea and Mauna Loa. You can look to the west to see the sun setting into the sea, its red disk disappearing and its green rim flashing momentarily.

In the twilight, you can imagine the people of old gathered here, listening to storytellers speaking of that land of Kahiki so far across the sea; hearing long, meandering stories of gods and grand-mannered chiefs, tales of capricious and insolent ones, and recitations of the exploits of the mischievous demigod Māui; accounts of great voyages and voyagers, the myths of the creation of earth and sky in the separation of Wākea from Papa, and epics of encounters with imagined creatures; stories of winds that bring messages in misty clouds from loved ones on other islands far over the sea; stories acted out and dramatized in song and dance. And we can imagine, too, those profound moments when genealogies were chanted letter perfect—the unwritten history of the people transmitted from one generation to another, affirming rootedness and identity.

A dismal coast? Haunting, perhaps: there is a great deal of feeling of presence here. You speak quietly, in hushed tones, as if aware of someone behind you saying, "Look carefully. Don't rush." Dry, dusty, open—a region of ancient habitation clinging to the side of a mountain, midway between its base on the dark ocean floor and its summit in the bright sky —no other place in Hawai'i is like it.

The land toward the west is even dustier and drier, although in ancient times before the introduction of cattle it was probably covered with vegetation. Birds sing, the wind blows gently, and the vegetation increases with altitude to the green upland forests. "The shores and sides of the hills had no indications of being inhabited," wrote Vancouver in March 1793, "and were almost destitute of vegetable productions." He described some cinder cones and recent flows, concluding, ". . . these, sofar as our glasses enabled us to distinguish, betokened this part of the island to have undergone some violent effects from volcanic eruptions."

The Luala'ilua Hills, 'twofold tranquility', rest on the

smooth slopes of Haleakalā at about 500 meters elevation. The road continues around the base of the group of hills at the lower edge. The climb to the top of Lualaʻilua Hills is easy, for the highest point is only 200 meters above the base. You can wander into old vents of the Hāna Volcanic Series here and explore the curvature of the cones.

Hōkūkano Cone (named for the star Hōkūpōkano) lies about 5 kilometers farther on, just seaward of the road. Crescent-shaped, Hōkūkano opens toward the ocean.

Puʻu Pimoe, 3 kilometers beyond Hōkūkano, is a cinder cone with washboardy sides. Pimoe, in legend, was a demigod who, in his *ulua* fish form, was hooked by Māui but escaped. Grass-covered, Puʻu Pimoe stands out in striking contrast to the black lava that poured from it. Highest on its western side, the asymetrical rim of Puʻu Pimoe sweeps and loops in front of the island of Kahoʻolawe across the channel.

Halfway to the sea at the edge of the Pimoe flow is the cone Pōhākea, 'white stone', at an altitude of 367 meters.

Lualaʻilua Hills. Erosion is beginning to cut gullies in the softer material at the base of the hill in the foreground.

Puʻu Pimoe.

The southwest rift and La Pérouse Bay. The dark lava in the foreground was the last active flow on Maui.

LA PÉROUSE BAY

Lava, flowing from two vents in recent times, poured into the sea to form La Pérouse Bay. The higher of the vents, elevation 472 meters, was active first, spilling lava down the slopes. Then a second, lower vent at elevation 175 meters opened a short time later. Lava from the two joined into a single flow 5 kilometers wide at the shoreline. Thirty million cubic meters of lava poured out onto the land, and an unknown amount flowed down into the depths of the ocean.

The eruptions at Keoneʻōʻio took place in about 1790, sometime between the visits of La Pérouse and Vancouver. La Pérouse, visiting the region in 1786, mapped a long shallow bay from Cape Hanamanioa to Puʻu Ōlaʻi. Vancouver's 1792 map shows a large peninsula of lava in the bay. The latest volcanic activity on Maui—and the latest in the Islands other than on Hawaiʻi—occurred, then, about two centuries ago.

The "King's Highway" runs across the Keoneʻōʻio flow. High chiefs and royalty walked this trail centuries ago; perhaps in spirit they still do. Two to three meters wide, the trail has a curbing of flat rock slabs standing on end marking one side of the path, and a line of boulders marking the other edge. Small red and white rocks wedged in between the large boulders add support and color.

Built of fitted boulders and chunks of cinder, the trail crosses jagged flows and bridges aa channels. A tremendous amount of energy must have been required to build it. Commitment, too—days of labor in the hot sun, shaping, leveling, bridging the path for kings. The trail is straight, for there is really no choice over a jagged lava field. No contouring is needed, only cut and fill. Pebbles and cobbles are loose on this road and it is a strain on the feet, since the rocks roll beneath the step, massaging the foot soles. It is a hot walk, as well; the black rock reradiates the sun's heat.

Beyond Cape Kīnaʻu, which was created by this most recent lava flow on the island, the shoreline swings northward

The King's Highway.

to the white sandy beach of ʻĀhihi Bay and the cone Puʻu Ōlaʻi. We saw another Puʻu Ōlaʻi on the windward side of West Maui near Kahakuloa; perhaps to avoid confusion, this one near Mākena is known locally as "Round Mountain."

Captain James King, leaving Kealakekua Bay after Cook's death there in February 1779, sailed along this coast of Maui. Describing this aspect of the island, he wrote in the ship's log that the mountainous parts

> are connected by a low, flat isthmus, appearing at first like two separate islands. This deception continued on the South West side, till we approached within eight or ten leagues of the coast, which, bending inward . . . formed a fine capacious bay. The Westernmost point . . . is made remarkable by a small hillock [Puʻu Ōlaʻi], to the Southward of which there is a fine sandy bay, with several huts on the shore, and a number of cocoanut trees growing about them.

When the sea was 8 meters higher, it notched Puʻu Ōlaʻi, and shells are found on that terrace. Crystals of olivine and augite

127

are common, some with arrowhead twins. The seaward side of the cone is steep; the landward side is covered with huge *kiawe* trees through whose tiny leaflets green light filters. The branches of the gnarled trees crack and bend to the ground, and, still attached to the trees, form arches.

Across ʻAlalākeiki Channel is Kahoʻolawe, in the rain shadow of Haleakalā, in ancient days an important island. Even though rainfall was sparse, wells made the island habitable. That delicately balanced ecosystem which then included man has since been severely disturbed with the introduction of goats as well as its use as a target area for practice bombing.

Puʻu Ōlaʻi lies at the southwestern corner of Haleakalā.

Time has been collapsed in this drawing to show three types of volcanic activity in the Maui group. In the background is the slowly built shield of Kahoʻolawe. At right, an offshore explosive steam eruption created the tuff cone now called Molokini Island. At left center is the fountaining cinder cone which became Puʻu Ōlaʻi. In the foreground, pahoehoe lava and rafted boulders move down the flank of Haleakalā.

From its 110-meter summit you can see eight great shield volcanoes: Mauna Kea, Mauna Loa, and Hualālai on the island of Hawaiʻi; Haleakalā and West Maui; Kahoʻolawe, Lānaʻi, and Molokaʻi—and Maui's only tuff cone, Molokini Island.

Molokini, 'many ties', rises from the sea 8 kilometers from Puʻu Ōlaʻi, its base on the southwest rift of Haleakalā. Molokini is crescent shaped, 25 meters high, and opens to the north. It is a tuff cone which came into existence in a series of steam explosions, as did Diamond Head on Oʻahu.

A tuff cone forms when molten rock, working its way toward the surface, superheats water trapped within porous rock. Pressure builds as that water reaches temperatures in excess of 100°C. Suddenly, the burden of overlying rock is exploded away as the water bursts into steam. With an instant thousandfold expansion of gas, the explosion blasts outward at low angle from the crater, hurling water, rock, ash, and dust into the air. Upon settling in a wide ring, the mound of ash and dust hardens into a buff-colored stone, volcanic **tuff.**

Molokini as it looks today, after eons of erosion. Note the wave patterns as currents swirl around the crescent.

THE SOUTHWEST RIFT

The 18-kilometer coastline from Pu'u Ōla'i to Keālia Pond runs almost directly northward. Spaced evenly along it at intervals of 3 kilometers are the villages of Mākena, Wailea, Keawakapu, Kama'ole, Lā'ie, and Kīhei. To the west are the islands of Lāna'i and Kaho'olawe, separated from one another by Kealaikahiki Channel.

Sandy beaches line the coast. Sunshine is plentiful, for this is the driest region on Haleakalā. Kīhei recieves less than 30 centimeters of rain each year, and most of that comes between November and April, peaking in January.

Near Kīhei, a fossiliferous marine conglomerate at an elevation of 15 meters indicates the extent to which Maui was flooded in times past during a higher stand of the sea.

Sharply dividing the western from the southern flanks of Haleakalā is the southwest rift, running from La Pérouse Bay to the summit. Westward from the rift the mountain slopes gently toward the Isthmus—a 6-degree slope, compared with

Southwest rift and summit of Haleakalā.

130

the 17 degrees on the southern side toward Kaupō. The west slope ends on the Isthmus, but the south slope extends far downward into the abyssal depths of the ocean.

Puʻu Māhoe cinder cone lies on this southwest rift, just above Piʻilani Highway, its top at an elevation of 811 meters. Imbedded within the cone are layers of soil. Usually, soil forms on the surface of a flow, and its thickness is some indication of the length of time between successive eruptions. Here, though, the soil was formed as water moved along certain layers, decomposing the rock. Two kilometers on each side of Puʻu Māhoe, in opposite directions, are the settlements of ʻUlupalakua to the north and Kanaio to the southeast.

An array of cones and craters marks the southwest rift. Volcanic eruptions up and down this region repeatedly blew ash and dust into the air which, upon settling, formed banks of material 2 to 3 meters thick near the vents and tapering to thin layers with increasing distance. Again and again, activity started at higher levels with lava fountains playing into the air. Then with the cessation of activity at the upper reaches, renewed volcanism would break out below. The higher regions, left unsupported, collapsed into pit craters. Such craters are prominent near the top of the rift.

A jeep road runs through pasture lands among big cones of the Hāna Volcanic Series. Old aa channels that once guided the lava to the sea are now grassy curving chutes with rock wall sides, intertwining in braids. A trail moves upward along the rift at an angle of 8 degrees, past 1600-meter Puʻu Mākua, and on into the Kula and Kahikinui Forest Reserve.

A loop trail at Polipoli, elevation about 1800 meters, circles through cypress, ash, redwood, and eucalyptus. It is profoundly quiet at Polipoli. Fog moves among the trees, alternately expanding and limiting what we can see. Sunlight filters through the forest crown to be diffused onto the soft floor 30 meters below.

Above Polipoli, the vegetation thins to occasional tufts of grass among black flows of lava. Explosions along the rift at

Craters and cones along the southwest rift above Polipoli.

about 2500 meters elevation showered the area with rocky blocks, some of them a meter or more in breadth. Near the summit are pit craters and cones of the earlier Kula Volcanic Series.

In *The Cruise of the Snark*, first published in 1911, Jack London described the illusion he experienced in his climb of Haleakalā:*

> There is a familiar and strange illusion experienced by all who climb isolated mountains. The higher one climbs, the more of the earth's surface becomes visible, and the effect of this is that the horizon seems uphill from the observer. This illusion is especially notable on Haleakala, for the old volcano rises directly from the sea, without buttresses or connecting ranges. In consequence, as fast as we climbed up the grim slope of Haleakala, still faster did Haleakala, ourselves, and all about us, sink down into the centre of what appeared a profound abyss. Everywhere, far above us, towered the horizon. The ocean sloped down from the horizon to us. The higher we climbed, the deeper did we seem to sink down, the farther above us shone the horizon, and steeper pitched the grade up to that horizontal line where sky and ocean met. It was weird and unreal, and vagrant thoughts of Simm's Hole and of the volcano through which Jules Verne journeyed to the centre of the earth flitted through one's mind.

*Reprinted in *The Spell of Hawaii* by Day and Stroven, 1968

Cones on the southwest rift near the Polipoli region. Aa channels snake along the
right edge of this photo.

THE SUMMIT

Red Hill, 3055 meters, is the highest point on Maui, and one
of three prominent cinder cones at the top of Haleakalā.

Pakao'ao cone is beside the Hawaii National Park Observatory at the crater's rim. It is capped with ankaramite, a
dense rock that was a favorite among ancient Hawaiians for
the making of stone adzes. Nearby is Magnetic Peak, named
for the effect it exerts on a compass needle. Lava bombs were
ejected from this vent; one of them is in the form of a sitting
duck.

Ankaramite, Pakao'ao cone.

A "science city" at the top of Haleakalā increases our sensitivity to the universe. Telescopes operated by the University of Hawai'i Institute for Astronomy watch the sun by day and the moon and stars by night. The Mees Observatory telescope uses a 30-centimeter mirror in photographing the mottled, granulated surface of the sun. It also looks at the edge of the solar disk, photographing the flares that leap high into space above the sun. Such observations lead to the development of stellar theory, to a clearer understanding of the burning of the sun, and to a knowledge of the evolution of stars in our galaxy as well as those in other galaxies in the universe.

Some nights you may see a brilliant green beam of light, startling in its brilliance, directed toward the moon. By the time the light of this laser beam reaches the moon, it has fanned out into a circle 3 kilometers in diameter. It searches out a reflector about 1 square meter in diameter, left on the moon by Apollo astronauts, then bounces back to the telescope on Haleakalā. The attenuation of the signal is tremendous. Of the 1,000,000,000,000,000,000 photons of light sent out in each burst, only between one and ten are captured by the telescope. Yet a great deal of information is contained in that feeble response—information, first, about the distance to the moon. The lunar distance is now known to within an accuracy of about 1 meter. Further refinement will increase that accuracy by a hundredfold, reducing uncertainty to the order of 1 centimeter.

Such precision helps determine the rate of movement of the big plates of the Earth's crust on which the continents and oceans ride. Combining these observations with data from other observatories that sense the moon with lasers, a baseline is established. A change in length of the line is a measure of plate drift.

A third experiment may help in the predicting of earthquakes. The Earth is constantly jiggling, quaking, shrinking, expanding, spewing out lavas, and cracking. The rotational poles change position slightly to accommodate for the shifting mass, and an observed polar shifting may presage earthquake activity. Formerly, the work of determining the change in rotational axis was done by using the stars. That process is complicated, and a year's calculation time followed the observations. Now, with the laser beam, the information may be generated quickly.

HALEAKALĀ CRATER

The summit of Haleakalā is an oval-shaped depression 12 kilometers long and 4 kilometers wide. The rim is broken in two places: Kaupō Gap, lying at the far eastern end of the depression, and Koʻolau Gap, opening into Keʻanae Valley on the northern side.

The floor of the crater, covered with colorful cinder cones, lies a thousand meters below the observatory rim. So deep is

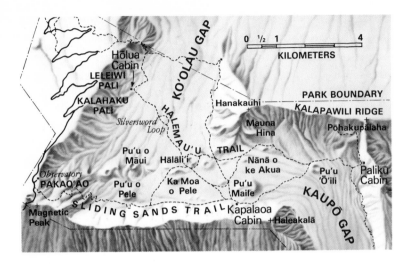

Map of Haleakalā Crater.

Haleakalā Crater from Kalahaku Pali. Left of center is Hanakauhi, the 'maker of mists.'

the crater and so far away are the cones that they look tiny when seen from the rim. Nothing is here to provide scale, except for an occasional group of hikers winding among the cones and kicking up a plume of dust that trails out behind them. Even Puʻu o Māui, a 300-meter giant, seems small in this great pit.

To the left of the observatory is Kalahaku Pali, and beyond it is Leleiwi Pali, directly above Hōlua Cabin. On the other side of Koʻolau Gap is Hanakauhi, 'maker of mists', 8 kilometers distant. Clouds move upward through the gap just this side of the mist-maker, pouring over irregular rivers of black lava and on to the dry crater floor. Here the clouds evaporate —disappearing without advancing—using energy just to stay in place.

Sometimes, with the crater filled with clouds and the sun behind you in the west, you can stand at the edge of the rim and see your own shadow projected onto the white mist below, magnified and encircled with a rainbow (a phenomenon known as the "specter of the Brocken").

Palikū Cabin is at the far end of the crater below a cliff; Kīpahulu Valley lies on the other side of it. Just this side of Kaupō Gap, at an elevation of 2500 meters, is the peak known as Haleakalā, 'house of the sun'. Stone shelters, platforms, and a *heiau* are here on the rim. Obviously this was an important place. Perhaps it was here that Māui snared the Sun.

Long ago, the Sun moved swiftly across the sky. The day was short. Night came too quickly—long before a day's work could be done. Far too little time was available for the goddess Hina to dry her fine tapa. The darkness of night was heavy, pressing down on the Earth with such force that it flattened the leaves on all the trees.

Hina's helpful, mischievous son, the demigod Māui, solved the vexing problem of diurnal brevity and nocturnal oppressiveness by slowing the Sun. First he made a net, using the fibers of the ʻieʻie vine and the *olonā* shrub. Then, net in hand, he waited for Lā, the Sun, to appear. From his hiding place he patiently watched the fading of the stars and the coming of morning twilight. Long, crepuscular rays crept higher and higher above the horizon, presaging the Sun's imminent arrival. And then Lā vaulted into the sky.

Timing his attack to just the right moment, Māui sprang and hurled the net high over the head of Lā, capturing him. Holding him firmly in the net, Māui refused to release him until he had extracted Lā's promise never to travel so fast across the sky again.

Now the day is longer than it used to be, and Māui, the half-mortal, again proved himself to be a benefactor of mankind. More time is now available for completing daylight tasks—fishing operations, the preparing of ceremonials, and the drying of tapa in the warmth of the Sun. Today the whole mountain, not just this peak, is known as the "House of the Sun."

Colors in the crater vary from red ochre to lavender and, in volcanic throats, from purple to gold. Color depends upon the chemical composition of the erupted lava and the degree of oxidation of the iron within it.

Jack London responded to the view with this description (again in *The Cruise of the Snark*):

And then, when at last we reached the summit of that monster mountain, which summit was like the bottom of an inverted cone situated in the centre of an awful cosmic pit, we found that we were at neither top nor bottom. Far above us was the heaven-towering horizon, and far beneath us, where the top of the mountain should have been, was a deeper deep, the great crater, the House of the Sun. Twenty-three miles around stretched the dizzy walls of the crater. We stood on the edge of the nearly vertical western wall, and the floor of the crater lay nearly half a mile beneath. This floor, broken by lava-flows and cinder-cones, was as red and fresh and uneroded as if it were but yesterday that the fires went out. The cinder-cones, the smallest over four hundred feet in height and the largest over nine hundred, seemed no more than puny little sand-hills, so mighty was the magnitude of the setting. Two gaps, thousands of feet deep, broke the rim of the crater, and through these Ukiukiu [a Haleakalā wind] vainly strove to drive his fleecy herds of trade-wind clouds. As fast as they advanced through the gaps, the heat of the crater dissipated them into thin air, and though they advanced always, they got nowhere.

But for all its magnificence, Haleakalā is not a crater, at least not a volcanic crater. A volcanic crater forms either as material is blown out of the top of a cone, or as a vent region sinks to become a caldera. Haleakalā is the result of neither explosive activity nor of sinking.

It happened this way: The Honomanū shield was completed, giving Haleakalā its general shape. Then lavas of the Kula Volcanic Series poured out onto that earlier shield, covering it unevenly and steepening its slopes. Kula volcanics are as much as 750 meters thick near the summit, tapering to as little as 15 meters at the shore. The summit was then a thousand meters higher than it is now and studded with cinder cones. At that time, Haleakalā may very well have had much the same appearance as Mauna Kea has today.

A long period of erosion followed Kula activity. Streams worked headward, enlarging and deepening the valleys, carving great amphitheaters with high waterfalls. The heads of Keʻanae and Kaupō Valleys came nearly to the point of joining. Not quite. A ridge down from the mist-maker Hana-kauhi is buried beneath the crater floor but still marks that old division. The huge amphitheaters of Kula times were doomed to extinction with the advent of Hāna activity.

A rift, opening across the mountain from Hāna to La Pé-rouse Bay, announced the arrival of Hāna times. Welling out along the rift, Hāna volcanics covered the earlier Kula lavas. Copious flows at the summit filled the region between the heads of Kaupō and Keʻanae to a depth of a thousand meters, producing a flat floor. The activity of Hāna times would be somewhat comparable to an imagined rift in the West Maui Mountains breaking open from Wailuku to Lahaina, flooding ʻĪao Valley to a high level, and pouring through gaps.

Haleakalā crater, then, is a summit depression that formed when lava flooded a region between the heads of two greatly eroded valleys.

One of the most rewarding and awesome experiences to be found in Hawai'i is to descend into the huge depression at the summit of Haleakalā and walk among its cones and pits, surrounded by the 37-kilometer rim, hundreds of meters beneath its jagged peaks, listening to its eternal silence—treading a landscape often and aptly decribed as lunar.

Keonehe'ehe'e Trail, 'sliding sands', begins between the cones of Pakao'ao and Magnetic Peak. We leave the road and suddenly plunge into a strange and alien world. A vast openness stretches before us as we hike down the dusty trail, suddenly isolated from the familiar world.

Distant cones grow in size as we descend. Spires, ridges, dikes, and jagged rocks surround us, along with fractures, pits, and craters. Tying it all together are long slopes of talus sweeping down from the summit in graceful catenary curves, covering old flows of aa and pahoehoe. There is scarcely any

Pu'u o Māui in Haleakalā Crater. Beyond the cone is the Sliding Sands trail.

Hikers circle a cinder cone.

vegetation at first, but plant life becomes more abundant the farther we go. Solitary clumps of grass begin to dot the region. Drifting, shifting sand piles up around the clumps and infrequent scrub. Rotating in the wind, branches of tethered plants bend to scrape concentric circles in the sand that sparkles in the sunlight. Ripple marks form and reform, and we measure the distances between crest and trough.

The trail makes a little jog to go around colorful Puʻu o Pele, one of the many abodes of the volcano goddess during her sojourn on Maui. Here we rest, empty the sand from our shoes, and, with greater appreciation, contemplate the word keoneheʻeheʻe. We look back to that towering rim from which we have made quite an easy descent in spite of the altitude. And we also look down to the vast expanse of crater floor still ahead, the trails clearly visible in the distance.

More and more plants appear as we near Kapalaoa Cabin. Mountain *pili*, a native bunchgrass, adds touches of green to the black sand. So, too, does the mountain *pilo*, a shrub with orange-colored berries belonging to the coffee family.

A gentle wind blows. Insects buzz in the silence. *Nēnē* geese squawk, mountain goats bleat, and cinders crunch beneath our feet. All else is quiet. Wisps of clouds bump noiselessly into the rocks of the cliff above. The sun is hot, the wind is cool. Silence and sound, hot and cold, pristine and primeval —fitting contrasts and contradictions in this unique experience.

High on the cliffs near the rim above us are hollows where the dark-rumped petrel builds its nest. Returning home from the sea each night at dusk, this bird emits a strange call somewhat like that of a small dog barking.

As we approach Kapalaoa Cabin, the trail is level and well defined, for it runs through a meadow of low green ground cover. Water, a stove, utensils, and bunks await us at the cabin—a fine place to camp. The grass is higher here. People in ancient times also camped near this place. Semicircular stone shelters, now alternately covered and uncovered with drifting sand, served as windbreaks for a large number of campers.

Beyond Kapalaoa, the trail crosses an old jagged aa flow that once moved down through Kaupō Gap. Sandalwood grows at lower levels in the gap—a tree recognized by its dark green, leathery leaves. During the summer season, deep red flowers appear at the end of its branches. The wood is fragrant and so prized that the tree was virtually exterminated a century ago in the era of sandalwood trade.

We leave a region of rough aa to cross a field of pahoehoe, a pavement solid to the foot and comfortable with its hard, smooth surface. Beside the trail, stunted *māmane* trees with tiny leaflets bloom yellow in the springtime. *'Ōhelo* berries, plentiful in their season, are delightfully edible. The jet black berries of the *kūkaenēnē* are a favorite of the *nēnē*; Hawaiians used this berry only as an emetic.

If it is misty here—and it often is—then it is even more misty as we hike toward Pu'u 'Ō'ili. Cairns, marking the trail like buoys, guide us through the encroaching fog. So we walk in clouds as we move toward that solitary cone, crossing con-

142

Exposed dikes at Kalahaku Pali.

gealed rivers of lava that once moved slowly across the crater floor, then accelerated as they poured down gaps in the mountainside.

Beyond Puʻu ʻŌʻili, the mist turns to the rains of Palikū, and we hear the waterfalls running off the ridge behind Palikū Cabin. The Lauʻulu Trail switches back and forth to ascend that ridge, climbing from 1950 to 2550 meters. From the top of the ridge we can look down into Kīpahulu Valley on a clear day, see the Hāna coast, and, in the distance, the great volcanic shields of the island of Hawaiʻi.

Several native trees still remain at Palikū: *ʻōhiʻa*, *ʻōlapa*, *kōlea*, and *manono*. But the meadow is now so thick with introduced grasses that seedlings from these trees have little chance of getting started. Young trees are not replacing older ones, and the forest is receding.

It is cold at night. Stars shine with a dazzling brilliance, for the altitude and the fact that most of the water vapor is

143

Haleakalā Crater from Kalahaku Pali. The farthest cinder cone *(center)* is Puʻu o Pele, and to its left is Puʻu o Māui. The two cones at right are Kamaʻoliʻi and Kaluʻu o ka ʻōʻō. Kaupō Gap is at extreme left, and in the distance are the shield volcanoes of Hawaiʻi Island.

frozen out of the atmosphere combine to make the skies wonderfully clear. The heavens take on a new dimension as stars which we have never seen before are revealed, so many that we must squint to allow the light from only the brightest ones through so that we can maintain our usual orientation.

Early on winter evenings, the "Canoe-Guiding Group," the Pleiades Cluster, passes overhead. This was one of the star groups that guided the Polynesians of old to Hawaiʻi. Those ancient voyagers called the group Na Huihui o Makaliʻi, 'the cluster of eyes'. It marked the northern limit on long journeys from Polynesia south of the equator. A line from the Pleiades toward the southeast passes through the three stars of Orion's belt and, continued, comes to the brightest star in the sky—Sirius, the star that passes over Tahiti.

Early in the evenings of late spring and early summer, orange-red Arcturus passes overhead. Hawaiians of old knew it as Hokuleʻa, 'the star of happiness'. A curving line through the handle of the Big Dipper sweeps through Arcturus and Spica, and continues far south to the Southern Cross. Hawai'i is the only island chain in Polynesia where you can see from North Star to Southern Cross.

A layer of ice covers the puddles beside the trail early on winter mornings at Palikū, and vegetation has a thin film of frost. But in the summertime, the evening primrose, which blooms at night with a sulfur-colored flower, turns red in the morning light.

The trail from Palikū Cabin parallels the face of the cliff at the north side of the crater. Kalapawili Ridge is the rim of the crater, 600 meters above the floor. The ridge itself runs from Hanakauhi to the head of Kīpahulu Valley, and along it are shelters and *heiau* platforms.

Pōhakupālaha, 'broad rock', is a high point on the ridge north of Palikū Cabin. All the boundary lines of the *ahupua'a* or ancient land divisions, of East Maui meet here, radiating from the point like spokes from the hub of a wheel . The *ahupua'a* of old extended from the uplands to the ocean, assuring the people living within them of access to both sea and mountain for sustenance.

Pu'u 'Ō'ili, a 100-meter cone, opens toward Palikū Cabin. Its twin, Pu'u Maile, opens in the opposite direction, toward Kapalaoa Cabin. From a point between the cones, we look down Kaupō Gap and across the sea to massive Mauna Kea, Mauna Loa, and Hualālai, rising mirage-like through long, streaked lines of trade wind clouds.

Slablike rock walls 8 meters wide jut out of the slopes below Hanakauhi. Light in color, they are among the widest dikes in all the islands. Erosion has partially uncovered these channels to ancient lava sources. Here too is the ancient divide between Kaupō and Ke'anae valleys, now deeply buried by Hāna volcanics.

Caves, pits, holes, and spatter vents—and even a highway —are found among the cones on the crater floor. Many of the features have picturesque names: Bubble Cave, Bottomless Pit, Pele's Pig Pen, Pele's Paint Pot.

Looking southeast from the rim of Haleakalā Crater, we see the great volcanic shields that make up the island of Hawaiʻi. *Left*, the Kohala Mountains all but hide Mauna Kea; *center*, the dome of Mauna Loa; *right*, Hualālai. To the extreme right on the crater rim is Kaupō Gap.

Pele's Paint Pot, a colorful region. Puʻu o Māui is the cone in the distance.

Bubble Cave is near Kapalaoa Cabin. Hot incandescent gases, blowing through molten rock, formed this large bubble with its heavy walls. The top of the bubble collapsed, affording access into the cave beneath. For centuries travelers have used Bubble Cave as a cozy protection from the icy winds of Haleakalā.

Bottomless Pit is a spatter vent with a low, colorful rim. The vent is 3 meters in diameter. At one time blobs of molten lava spewed out of it. The relatively little spatter present indicates that the vent served chiefly for the discharge of superheated gases. The pit is deep, but not unfathomable. The sides of this conduit are vertical and it bottoms out at 30 meters. Hawaiians of old placed umbilical cords in it to insure safe protection and growing strength for the newborn child.

Puʻu Nole, 'grumbling hill', is young—so recent that it is still black. Its name suggests that it was active until very recent times. The nearby cones of Mauna Hina and Nānā o ke Akua are much older eruptions, now covered with shrubs and grass.

Ka Puaʻa o Pele, 'Pele's pig pen', was a place of high *kapu* in ancient times. It is another spatter vent, 10 meters square, opening to the northwest. The enclosure is surprisingly rectangular; its walls are 2 to 3 meters high. Partially buried, the vent lies in the saddle between the slopes of the cinder cones Ka Moa o Pele and Hālāliʻi.

On the opposite side of Hālāliʻi, the trail winds and loops in the colorful region of Pele's Paint Pot. Long, clear, sharp slopes of varicolored cones interweave, and the trail meanders through the maze, rising, curving, and falling. Through the gap ahead is Puʻu o Māui, and far in the distance at the rim of the crater is the observatory.

The trail divides many times as it wanders about the crater, circling the bases of the cones. So there are many choices, and all are right. We are confronted only with one consideration: "What can we leave unexplored?"

Silversword is startling to see, so white and silvery in black

Pele's Pig Pen, a remnant of a spatter cone.

sand, the stiletto-shaped leaves pointing upward from the rounded mass of the plant. It has adapted well to its habitat, flourishing where nothing else does. Before the introduction of goats to the island, silverswords must have dotted the entire summit of Haleakalā.

Conditions in the crater are rugged for any plant—hot in the day, cold at night. Humidity is practically zero, and rainfall is slight. Silversword not only survives, it thrives under the conditions we deem hostile. Buried deep within the plant's protective rosette is a growing center. The silvery hairs on its leaves protect against excessive sunshine. Flat, concave, and numerous, the hairs reflect light, giving the plant the patina expressed in its Hawaiian name, ʻāhinahina, 'very gray'. Narrow leaves absorb sunshine, and small surface minimizes evaporation. During long periods between rains, water is stored in gelatinous accumulations within the leaves.

Silversword is a slow bloomer, taking from four to twenty years to flower. Blossoming during the summer on stalks a meter or two high, it produces a hundred or more purplish blooms like tiny sunflowers. After the seeds mature, the plant dies.

Haleakalā, rising 3.1 kilometers above the sea, would seem to present a formidable barrier to communication among the groups of Hawaiians living here in times past. With that assumption in mind, it is a surprise to come across a highway at this elevation. Built by Kihapiʻilani, the highway runs toward Hālāliʻi. Only parts of it remain visible now, as it has been almost completely covered with drifting sand. A line of communication between Wailuku and Hāna, this route may have been preferred, regardless of its hardships, to routes at lower levels through rain forests, over gulches, and across deep valleys.

Small piles of stone—*ahu*—served the ancients as guides to

Silversword in bloom.

the trails through the mists. Many of the *ahu*, stone enclosures, altars, and terraced platforms mapped by Dr. Emory in 1922 have since disappeared beneath drifting sand.

The trail toward Hōlua Cabin crosses a flow of black lava that came from the central cluster of cones. Farther along, it crosses a reddish lava flow that came from Puʻu o Māui. The red changes to black again on a flow from Kamaʻoliʻi. Here, on an old aa surface, the trail is anything but smooth.

Beside the trail just a few hundred meters from Hōlua Cabin is Na Piko Haua, 'hiding place of navel cords'. It is a spatter vent 5 meters in diameter, similar to Bottomless Pit. Umbilical cords, wrapped in tapa and tied with fiber cord, were hidden in rock crevices that were then sealed with wedged stones.

The green meadows and tall grass at Hōlua Cabin are a welcome sight after the roughness of the aa trail. It is a place to stretch and relax, in touch with the cool verdure, and think about the ascent of the wall of the crater. Above the cabin is Hōlua Cave, which served as a shelter until the cabin was built. The evidence from radiocarbon dating of the residue from ancient fires indicates that the cave must have been in use nearly a thousand years ago.

Halemauʻu Trail parallels the thousand-meter Leleiwi Pali for 1.5 kilometers, then begins to climb the crater wall in a series of astounding switchbacks. A slow pace takes us easily around turn after turn, wandering in and out of gulches— scaling the mountainside, gaining altitude sometimes rapidly, always gradually.

As the switchbacks carry us back and forth across the face of the Pali, great vistas continually change. When mists roll in through Koʻolau Gap, we're surrounded and suddenly isolated, hanging on the edge of a cliff with only clouds below.

The clouds sweep away, revealing black and brown lavas that once pushed down to the sea at Keʻanae, and disclosing Hanakauhi, the maker of the mists. We look far across the crater toward Palikū and Kapalaoa cabins, nestled at the foot of crater walls, and their associated cones Puʻu ʻŌʻili and

Puʻu Maile. We see the cones in the central cluster that we have visited: Puʻu o Pele, Ka Moa o Pele, Puʻu o Māui, and the home of Pele's brother, Kamohoaliʻi, the king of vapor. Closest to us is the cone Kaluʻu o ka ʻōʻō. We see the long talus slopes, sweeping down to the crater floor. And we see the trail, think of the rocks, the rippled sand, and the footprints we made, so soon to be erased by the wind.

Along the trail are *amaʻumaʻu* ferns, some of which have reddish fronds. The young shoots were eaten by Hawaiians, the fronds used for thatching. Halemauʻu, 'grass house', is the name of this trail, built in 1937. It is named for a house with thatched roof that once stood at the rim as a shelter for travelers.

Finally the trail breaks out of the crater and heads downslope to Hosmer's Grove. Gone is the exciting panorama of crater vistas. Upslope behind us is the alpine summit of Haleakalā, bare and rocky. But it is not devoid of living things, for here grow lichens, moss, bentgrass, and silversword.

A succession of climatic zones lies beneath us, expressed in *māmane, naio, pukiawe, pilo, ʻōhelo*, fern, guava, lantana, *pānini*, and *kiawe*—regions of rain forest, grassy meadows of grazing land, broadlands of pineapple and sugarcane, and, finally, the warmth of the desert Isthmus.

Once this island called Maui, 4000 kilometers from the nearest continent, was without vegetation. But the dispersal of tiny spores on the wind, so tiny that they are measured in millionths of a meter, brought ferns, mosses, fungi, algae, and lichens. Birds found the island, bringing with them seeds stuck to their feathers or lodged within their digestive systems. Rafting and flotation may account for the presence of *hala* trees near the shore.

A delicately balanced ecosystem developed over a period of hundreds of thousands of years—a system defined by the ocean barrier. New arrivals upset the equilibrium, forcing changes. The system adjusted, and a new equilibrium came into being. With the arrival of man, the tendency toward instability increased. But with man also came the possibility of

Looking northwest from the summit of Haleakalā. The near cone is
Pu'u Nianiau; beyond is Hakuhe'e Point, West Maui, and the island of
Moloka'i is in the distance.

restoration through intelligent manipulation of the ecosystem.

The island ecosystem is ultimately destined to extinction.
So, too, is the island itself, but on a time scale of such magnitude that it holds little meaning for us. For the islands are
riding a moving crustal plate, traveling inexorably toward
trenches at the edges of the Pacific, heading for an ultimate
reincorporation into the innards of the Earth.

Long before Maui reaches its trench, though, waves and
rains will have reduced it to a shoal in the sea.

That is in the future.

But what of the present—what of this island that appeared
late in the geological history of the earth? In our lifetime, it
will move only as far to the northwest as we are tall. Thousands of human lifetimes yet lie ahead. And, in that perspective, the mountains of Maui are timeless and eternally serene.

References

Bier, J. A. *Map of Maui*. 2nd ed., rev. Honolulu: University Press of Hawaii, 1979.

Carlquist, S. *Hawaii: A Natural History*. Garden City: The Natural History Press, 1970.

Cook, J., and J. King. *A Voyage to the Pacific Ocean, 1776–1780*. Vols. 2 and 3. London: H. Hughs, 1784.

Cox, J. H., and E. Stasack. *Hawaiian Petroglyphs*. Honolulu: Bishop Museum Press, 1970.

Emory, K. P. An Archaeological Survey of Haleakala. *B. P. Bishop Museum Occasional Papers* 7(11):237–259. Honolulu: Bishop Museum Press, 1921.

Kay, A. *A Natural History of the Hawaiian Islands*. Honolulu: University Press of Hawaii, 1972.

Kyselka, W., and R. Lanterman. *North Star to Southern Cross*. Honolulu: University Press of Hawaii, 1976.

La Pérouse, J. F. G. de. *A Voyage Round the World Performed in the Years 1785, 1786, 1787, and 1788*. London: A. Hamilton, 1799.

London, J. *The Cruise of the "Snark."* New York: Macmillan, 1911. Excerpts reprinted in A. Grove Day and Carl Stroven, eds., *The Spell of Hawaii*, Meredith Press, 1968.

Macdonald, G. A. *Volcanoes.* New Jersey: Prentice-Hall, 1972.

Macdonald, G. A., and A. Abbott. *Volcanoes in the Sea.* Honolulu: University of Hawaii Press, 1970.

Macdonald, G. A., and D. H. Hubbard. *Volcanoes in the National Parks in Hawaii.* Hawaii Volcanoes National Park: Natural History Association, 1973.

Macdonald, G. A., and W. Kyselka. *Anatomy of an Island.* Honolulu: Bishop Museum Press, 1967.

Manapau, T. K. "A Visit to Kaupo." *Kuokoa,* June 15, 1922. Translation from manuscript verified by K. P. Emory, Department of Anthropology, B. P. Bishop Museum.

Pukui, M. K., S. Elbert, and E. Mookini. *Place Names of Hawaii.* Honolulu: University Press of Hawaii, 1974.

Soehren, L. J. "An Archaeological Survey of Portions of East Maui, Hawaii." Mimeographed. Prepared by Bishop Museum for the U.S. National Parks Service.

Stearns, H. T., and G. A. Macdonald. *Geology and Ground-water Resources of the Island of Maui, Hawaii.* Honolulu: Hawaii Division of Hydrography, Bulletin 7, 1942.

Vancouver, G. *A Voyage of Discovery to the North Pacific Ocean, and Round the World . . . 1790–1795.* London: 1798.

Walker, W. M. "Archaeology of Maui." Department of Anthropology, B. P. Bishop Museum, 1931 (unpublished manuscript).

Wenkam, R. *Maui: The Last Hawaiian Place.* San Francisco: Friends of the Earth, 1970.

Index

Aa. *See* Lava
'Alalākeiki: cave, 119; channel, 6, 9, 119, 128
'Ālau Islet, 105
'Alenuihāhā Channel, 8, 9, 112, 122
Alexander, Samuel T., 87
Alluvial fans, 28, 46, 47, 48, 51, 53
Alluvial plain, 28, 36, 72
Alluvial ridge, 36
Alluvium, 28, 30, 74, 92
Amphitheaters, 32, 36, 50, 62, 71, 72, 92, 93, 94, 99, 111, 114, 139
Anakaluahine Gulch, 63
Andesite, 44, 118
Ash, volcanic, 6, 16, 45, 56, 62, 63, 104, 105, 129, 131

Baldwin, Henry P., 87–88
Basalt, 20, 27, 51, 61, 64, 70, 89, 92, 118
Basaltic columns, 100, 109, 110
Bay mouth bar, 78, 79
Beach rock, 56
Bishop Museum, 76, 94, 115
Bloodstone at Ke'anae, 94
Bog, peat, 69

Bottomless Pit, 145, 147, 150
Bubble Cave, 145, 147

Caldera of West Maui volcano, 20, 32, 34, 36
Caves, 39, 100, 101, 119, 145, 147
Cinder, 16, 18. *See also* Cone, cinder; Ash
Clay, 30, 31, 34, 69
Composite volcano, 16–18 passim
Cone, cinder, 22, 37, 86, 88, 102, 131, 140–152 passim. *See also* Pu'u Hele; Pu'u Keka'a; Pu'u Ōla'i
Cook, Captain James, 28, 75, 84, 120, 127
Coral, 6, 41, 49, 56, 60. *See also* Reef
Crossbedding, 75

Diamond Head, 6, 129
Dikes, lava, 16, 17, 32, 36, 45, 46
Dunes, 27, 71, 74, 75, 76

Earth: crust of, 12, 16, 19, 135, 152; mantle of, 19. *See also* Hot Spot; Pacific Plate
Emperor Seamounts, 19

Fishpond, 105
Fissures, 16, 17, 62
Fleming Bay and beach, 57

Glaciation, 74. *See also* Ice age

Haʻikū (plantation and town), 88
Haipuaʻena: falls, 92; stream, 91
Hakuheʻe Point, 28, 70
Hālāliʻi, 147, 149
Haleakalā, 4, 5, 8, 10, 27, 69, 74, 78,
    83–152; emergence and building of,
    20, 22; former home of Pele, 53;
    peak (House of the Sun), 83, 137;
    sediments from, 80
Halemauʻu Trail, 150, 151
Hāmākua Ditch. *See* Alexander; Bald-
    win; Irrigation
Hāna Bay, 103, 104
Hāna (district, town), 22, 76, 85,
    102–108, 112, 115, 139, 143, 149
Hanakaʻōʻō Point, 55
Hanakauhi, 137, 139, 145, 150
Hāna Volcanic Series, 22, 86, 92, 94,
    96, 99, 104, 108, 111, 112, 118,
    131, 139
Hanawī (gulch, stream), 95
Hawaiian Chain, 3–5, 15, 19
Hawaiian Ridge, 8, 18, 19, 20
Hawaiian Swell, 19
Hāwea Point, 57, 58
Heiau, 52, 76–78; Halekiʻi, 27, 72,
    76–78; Halekumukalani, 51; Hale o
    Kane, 114, 117; Hale o Lono, 104;
    Halulukoakoa, 52; Heakalani, 62;
    Kaluanui, 107; Loʻaloʻa, 114;
    Pihana, 27, 72, 76–78; Piʻilanihale,
    99; Popoiwi, 113, 114; Wailehua,
    51, 52
Hoalua (bay, stream), 89, 90
Hōkūkano Cone, 124
Hōkūʻula, 107
Hōlua (cabin, cave), 137, 150
Hono-a-piʻilani Highway, 56
Honokahua (bay, village), 58–59
Honokalā Point, 89
Honokeana Bay, 56, 57
Honokōhau: bay, 62; district, 64;
    stream, 71, 72
Honokōwai (bay, point), 56, 57
Honolua Bay, 60, 61
Honolua Volcanic Series, 20, 50, 58,
    60, 61, 62, 63, 70

Honomāʻele, 99
Honomanū (bay, district, valley), 91,
    92, 95
Honomanū Volcanic Series, 22, 87,
    91, 92, 111, 139
Hononana, 65
Hoʻolawa, 89
Hosmer's Grove, 151
Hot Spot, 19
Huelo Point, 89

ʻĪao: needle, 29; stream, 27, 28,
    29–36, 72; valley, 29–36, 48, 50,
    76, 139
Ice age, 43
Irrigation, 37, 47, 51, 53, 54; ancient,
    at Kaupō, 120; Hāmākua system,
    87–88; of Isthmus, 82, 87. *See also*
    Haʻiku
Isthmus, 5, 10, 22, 36, 37, 71, 73–82,
    85, 87, 130, 131, 151

Jointing planes, 31, 41, 110

Kahakuloa, 65–68, 70, 127
Kahekili, 29, 76, 104, 117
Kahikinui, 121–123
Kahoma Stream, 53
Kahului Bay, 10, 37, 70, 78; break-
    water in, 26, 53, 76; reef at, 85;
    visited by Cook, 28
Kaʻīlio, 117
Kaʻinalimu Bay, 102, 103
Kaiwi Channel, 8, 9
Ka Iwi o Pele, 105, 107
Kalahaku Pali, 137
Kalanikupule, 29, 30, 48
Kalaniʻōpuʻu, 76, 104, 117
Kalohi Channel, 8, 9
Kaluʻu o ka ʻōʻō, 151
Kamaʻoliʻi, 150
Kamehameha, 29, 30, 48, 76, 104,
    112, 117
Ka Moa o Pele, 147, 151
Kanahā Pond, 26, 76, 78–79
Kanahā Stream, 53
Kanaio, 131
Kanounou Point, 63
Kapalaoa Cabin, 141, 142, 145, 147,
    150
Ka Puaʻa o Pele. *See* Pele's Pig Pen
Kapuaʻi o Kamehameha, 89
Kauaʻula Valley, 50–51, 53

Ka'uiki Head, 104
Kaupō (district, valley), 104, 112, 114, 115–120, 131
Kaupō Gap, 135, 142, 145
Kealaikahiki Channel, 9, 10, 130
Kealaloloa Ridge, 43
Keālia Pond, 79–80, 130
Keali'i Point, 89
Ke'anae (district, valley), 92, 93–94, 97, 135, 145, 150
Keawalua Gulch, 63
Keonehe'ehe'e Trail, 140
Keone'ō'io, 126
Keōpuka Rock, 91
Kepaniwai, 29, 30
Kiakeana Point, 120
Kīhei, 130
Kīlea cinder cone, 48, 49
King, Captain James, 127
King's Highway, 126
Kīpahulu (district, valley), 85, 96, 105, 109–112, 119, 120, 137, 143, 145
Ko'olau Gap, 94, 135, 137, 150
Kūhiwa (stream, valley), 95, 98, 99
Kuiki Peak, 113
Kukui'ula, 112
Kulaokalālāloa, 63
Kula Volcanic Series, 22, 86, 88, 92, 96, 109, 111, 112, 113, 118, 132, 139

Lahaina, 43, 51–53, 112, 139
Lahaina Volcanic Series, 20, 48, 53, 55
Lā'ie, 130
La Pérouse, J. F. G. de, 120, 126
La Pérouse Bay, 22, 126, 130, 139
Launiupoko Canyon, 49–50, 51, 53, 70
Lau'ulu Trail, 143
Lava: aa, 22, 80, 131, 140, 142, 150; ejecta, 37, 55, 133, 147; pahoehoe, 22, 140, 142; plastering, 94, 97; ponding of, 22, 74, 100; viscosity of, 15, 16, 41, 43, 46. See also Ash, volcanic; Basalt; Caves; Dikes, lava; Tuff cones
Lava tubes, 45, 100, 101
Leleiwi Pali, 137, 150
Līpoa Point, 58, 60
Lithification of sand dunes, 75, 76
Luala'ilua Hills, 123, 124

Mā'alaea Bay, 10, 37, 78
Magma, 12, 16, 17, 32, 61
Magnetic Peak, 133, 140
Makaīwa Bay, 91
Makāluapuna Point, 58, 59
Makawao, 87, 88
Mākena, 127, 130
Makuleia Bay, 60
Māla (settlement, wharf), 53, 54
Māliko (bay, gulch, stream), 86–88
Mamalu Bay, 117
Manapau, Thomas, 115–119 passim
Manawainui Canyon (East Maui), 113, 114, 117
Manawainui Gulch (West Maui), 43, 45
Māui (demigod), 123, 124; legends of, 105–107, 137–138
Maui-nui, 9
McGregor Point, 28, 39–43
Mōiki Point, 91
Mōkae Cove, 108
Mōke'ehia islet, 70
Mokulau, 112–114, 120
Moloka'i Channel, 8
Molokini Island, 6, 129
Moonstones, 34
Mount Ball. See Pa'upa'u
Mount 'Eke, 63, 68–70
Mudslide at Kaupō, 117–118
Mugearite, 20
Mū'olea Point, 109

Nāhiku, 95–98
Nākālele Point, 63
Nāmalu Bay, 57
Nānā o ke Akua, 147
Na Piko Haua, 150
Nāpili Bay, 57, 59
Nehe Point, 72
Nu'u, 119, 120

Observatory, 133, 137, 147
'Ohe'o Gulch, 110
Olinda, 88
Olivine, 127
Olowalu (canyon, stream), 30, 47–49, 70
Oneloa Bay, 58
O'opuola Stream, 91
'Ōpana Point, 88, 89

Pacific Plate, 19, 135

Pahoehoe. *See* Lava
Pāʻia and Lower Pāʻia, 85–86
Paʻiloa Bay, 100
Pailolo Channel, 9, 59, 62
Pakaoʻao cone, 133, 140
Palikea Stream, 110, 112
Palikū Cabin, 113, 117, 137, 143, 145, 150
Papahawahawa Stream, 109
Papiha Point, 95
Paukūkalo, 27, 72
Paʻupaʻu (Mount Ball), 51, 53
Pauwalu Point, 94
Paʻuwela Point, 85, 86, 88
Pele's Paint Pot, 145, 147
Pele's Pig Pen, 145, 147
Petroglyphs, 48, 109, 119
Pictographs, 108–109
Piʻilani: bays of, 56; highway, 131; trail, 99
Pīlale Bay, 89
Pit crater, 132
Poʻelua Beach, 65
Pōhākea Cone, 124
Pōhakukani 'bell stone', 65
Pōhaku koko 'bloodstone', 94
Pōhakupālaha, 145
Polipoli, 131
Pūehuehu Nui Gulch, 50, 53
Pumice, 38, 63
Puohokamoa Bay, 91, 92
Puʻu Anu, 43
Puʻu Hele, 37–39
Puʻu Hipa, 50
Puʻu Ka Ae, 89
Puʻu Kaʻeo, 63
Puʻu Kekaʻa, 55
Puʻukiʻi, 104
Puʻu Koaʻe, 67, 68
Puʻu Koʻai, 46
Puʻu Kukui, 28, 53, 62, 68, 71
Puʻu Laina, 53
Puʻu Lūʻau, 43
Puʻu Māhanaluanui, 49, 50
Puʻu Makawana, 71
Puʻu Mākua, 131
Puʻu Maneʻoneʻo, 117, 118
Puʻu ʻŌʻili, 142, 143, 145, 150
Puʻu Ōlaʻi, East Maui, 126–129 passim
Puʻu Ōlaʻi, West Maui, 28, 71, 127
Puʻu o Māui, 137, 147, 150, 151
Puʻu o Pele, 141, 151

Puʻu o ʻUmi, 87
Puʻu Pimoe, 124

Rainfall: on Mount ʻEke, 69; on Hale-akalā, 81; at Kaupō, 120; at Kīhei, 130; at Kīpahulu, 112; at Kūhiwa, 99; on Puʻu Kukui, 68; at Olinda, 88
Rains, Hawaiian names for, 116
Red Hill, 133
Reef, 6, 71, 85–86. *See also* Coral Rift: east, 22; north, 22, 86; southwest, 22, 130–132, 139
Rift zone, 16, 104

Salt pans, 119
Science City, 134
Sea caves, 63, 71, 94
Sea cliffs, 7, 91; at Honokōhau, 62; at Kahakuloa, 65, 70, 71; at Kaupō, 112; at Kīpahulu, 110; at McGregor Point, 41, 43; near Nāhiku, 95, 96; near Nuaʻailua, 93; at Puʻu Mane-ʻoneʻo, 117; at ʻUlaʻino, 99
Sea stacks: at Hoalua Bay, 90; at Ka-nounou Point, 63; at ʻOpana Point, 89; at Pauwalu Point, 94; at Puoho-kamoa Bay, 91; at Wailuaiki Bay, 95
Seven Pools of Pīpīwai Stream, 109, 110
Shells, 49, 66, 117, 127
Shield volcano, 15–18, 20, 32; of Haleakalā, 84, 93
Silt, 31, 32, 85
Silversword, 69, 147–149, 151
Sinkholes, 69, 70
Sliding Sands Trail, 140–142
Specter of the Brocken, 137
Spreckels, Claus, 87
Stands of the sea: high, 7, 9, 29, 41, 49, 50, 55, 90, 94, 127, 130; low, 9, 37, 43, 74, 94, 118
Stone shelters: at Kaupō, 116; in crater of Haleakalā, 137, 142
Stream piracy, 71, 72

Tahiti, 10, 60, 94, 103, 144
Talus, 151
Terracing, 29, 64, 112
Trachyte, 20, 50, 51, 58, 60, 61, 62, 69, 70
Trade winds, 67, 85, 105

Tuff cones, 6, 7, 129

Ukumehame Canyon, 16, 45–47, 70
ʻUlaʻino, 99
ʻUlupalakua, 131

Vancouver, Captain G., 123, 126

Waiʻānapanapa, 101
Waiehu Stream, 28
Waiheʻe (district, point, stream), 27,
    28, 64, 71, 72, 78. *See also* Dunes
Waihoʻi Valley, 108, 109, 110, 112
Waikamoi, 91
Waikapū (stream, valley), 28, 36–37,
    81

Wailea, 130
Wailua: bay, 95; cove, 109; district,
    109–110, 112; village, 94; water-
    fall, 110
Wailuku, 28, 43, 75, 99, 139, 149
Wailuku Volcanic Series, 20, 61, 63
Waiohonu Stream, 108
Waiokamilo Stream, 95
Waipiʻo Bay, 89, 91
Waiū, 117
Weathering, 31, 66
Wells, 80–82
West Maui, 5, 16, 22, 74, 84, 127,
    129; emergence and building of,
    20–21; features of, 24–72
Winds, Hawaiian names for, 115–116

159

⚇ Production Notes

This book was designed by Ray Lanterman and
typeset on the Unified Composing System by the
design and production staff of The University Press
of Hawaii.

The text typeface is Compugraphic Caledonia
and the display typeface is Univers.

Offset presswork and binding were done by Halliday
Lithograph. Text paper is Glatco Matte, smooth
finish, basis 60.